The Biology of Tardigrades

QL	125554
447.5	Kinchin, Ian M.
.K55	The biology of
1994	Tarigrades

The Biology of Tardigrades

The Biology of Tardigrades

By Ian M. Kinchin

PORTLAND PRESS

Published by Portland Press Ltd.

Portland Press Ltd
59 Portland Place,
London W1N 3AJ
U.K.

In North America orders should be sent to

Portland Press Inc.
P.O. Box 2191
Chapel Hill
NC 27515-2191
U.S.A.

Copyright © 1994 Portland Press, London

All rights reserved. Apart from any fair dealing for the purposes of research or private study, or criticism or review, as permitted under the Copyright, Designs and Patents Act, 1988, this publication may be reproduced, stored or transmitted, in any forms or by any means, only with the prior permission in writing of the publishers, or in the case of reprographic reproduction in accordance with the terms of licences issued by the Copyright Licensing Agency. Inquiries concerning reproduction outside those terms should be sent to the publishers at the above-mentioned address.

Although, at the time of going to press, the information contained in this publication is believed to be correct, neither the author nor the editors nor the publisher assume any responsibility for any errors or omissions herein contained. Opinions expressed in this book are those of the author and are not necessarily held by the editors or the publishers.

British Library Cataloguing in Publication Data
A catalogue record for this book is available from the British Library

ISBN 1-85578-043-7

Printed in Great Britain by the University Press, Cambridge

Dedication

To all the biologists at Westfield College, London (1980 – 83) particularly Jim Green, Marcus Trett and Tony Wallwork.

"I do not know what I may appear to the world, but to myself I seem to have been only a boy playing on the seashore, and diverting myself in now and then finding a smoother pebble or a prettier shell than ordinary, while the great ocean of truth lay all undiscovered before me."

Sir Isaac Newton.

Contents

Preface xi

1 Introduction

What is a tardigrade? 1
Historical perspective 2
Aims of this book 4

2 Origins and systematics

Fossil record 7
Affinities of the Tardigrada 8
Systematics of the phylum 12

3 External morphology

Morphological groups 25
Arthrotardigrada 25
Echiniscoidea 28
Eutardigrada 33

4 Internal structure

Internal arrangement 39
Intestine 39
Cuticle 45
Muscle 48
Nervous system and sensory organs 50
Excretory organs 52
Body cavity cells 54

5 Reproduction and life history

Mating 57
Hermaphroditism 57
Reproductive organs 58
Reproductive modes 60

Parthenogenesis 61
Spermatozoa 62
Eggs and Embryos 62
Hatching and post-embryonic development 66
Moulting 67
Cyclomorphosis 72

6 Cryptobiosis

The need for latent states 75
Anhydrobiosis 76
Cryobiosis 82
Osmobiosis 83
Anoxybiosis 83
Encystment 84

7 Ecology

Limno-terrestrial habitats 87
Marine habitats 90
Habitat preference 93
Distribution and dispersal 97
Population dynamics and densities 99
Associated microbes 103
Associated microfauna and trophic relationships 106

8 Collecting and preserving

Collecting 115
Culturing 116
Mounting 116

9 Guide to common species

Introduction 119
Quantitative morphometric data 119
Macrobiotus cf. *hufelandi* 123
Macrobiotus cf. *harmsworthi* 124
Milnesium tardigradum 126
Ramazzottius cf. *oberhauseri* 128
Echiniscus testudo 129
Echiniscus granulatus 129

Hypsibius dujardini 130
Hypsibius cf. *antarcticus/arcticus* 130
Diphascon scoticum 131
Mesocrista spitsbergense 133
Richtersius coronifer 133
Dactylobiotus dispar 134
Minibiotus intermedius 134
Batillipes spp. 135
Echiniscoides cf. *sigismundi* 135

10 Future research

The scope for further research 139
Behaviour 139
Biochemistry 141
Ecology 142
Reproductive strategies 144
Structure 144
The British fauna 145

Acknowledgements 147

Glossary 149

References 163

Subject index 183

Preface

My interest in tardigrades was fired by the enthusiasm of one of my teachers when I was an undergraduate. Like many zoologists, he was fascinated by these animals, but did not have the time to study them in detail. Once I had spent some time examining tardigrades, I was hooked by these engaging animals. Their ability to suspend metabolism, the doubts surrounding their phylogenetic relationships, their amusing gait and the ease with which the animals could be found in local microhabitats all added to their attractiveness as a subject for further study.

However, when the interested observer starts to scan the available literature concerning this group, it will be found that the accounts in many general zoology texts only whet the appetite and provide more questions than answers. Much of the specialist literature is scattered in various (often obscure) journals and published in a number of languages. This book (concentrating on recent developments whenever possible) provides the only comprehensive account of marine and freshwater tardigrade biology to be published in English and the first taxonomic review (in any language) for over a decade.

I hope that this book will be of interest not only to those studying tardigrades directly, but also to those who come into contact with tardigrades on a casual basis when studying marine or freshwater meiofauna. The wide variety of habitats colonized by tardigrades means that many biologists will encounter tardigrades from time to time. Armed with a modern synthesis such as this, the student of tardigradology can hopefully spend more time observing the animals and less searching for references in the library.

I.M.Kinchin
January 1994

Introduction

What is a tardigrade?

One of the factors which has contributed to the interest biologists have shown in tardigrades is the uncertainty surrounding their systematic position. Tardigrades exhibit an odd combination of traits, shared with such diverse groups as deuterostomes, pseudocoelomates and coelomate protostomes. Many biology books avoid the problem of tardigrade classification by simply describing them as a phylum of uncertain affinities. Unfortunately, this situation can lead to confusion as observations on tardigrades end up dispersed throughout the literature in unpredictable places. Therefore, for the convenience of future students of tardigradology, it is desirable to allocate a position to the Tardigrada within the invertebrates.

Often described as a 'minor phylum', tardigrades are better described as one of the 'lesser-known' phyla (Nelson, 1991) whose relationship with other groups is unclear (see Chapter 2). Such *Problematica* have been described as one of the most intriguing, and most ignored, of the problems in biology. The tendency of relegating them to the 'sidelines of enquiry' is understandable, but threatens to remove an area of great interest to evolutionary biology (Simonetta and Conway Morris, 1991).

Tardigrades are all aquatic animals, requiring a film of water around their bodies to permit locomotion and gas exchange. However, many species are typical of wet terrestrial habitats, such as moss cushions, from which they are most commonly found. These are described throughout this book as limno-terrestrial species. The separation of limno-terrestrial species from the truly freshwater species (from ponds and streams) is not always clear — with some species found in both types of habitat. The term limno-terrestrial has, therefore, been used to describe all these species here and only distinguishes them from the marine species.

The animals are typically in the range of 100–500 μm in length and so are visible only under the microscope. A detailed account of tardigrade morphology is given in Chapter 3 — the body is in the form of a cylinder, flattened on the underside with four pairs of clawed legs occurring ventro-laterally (Figure 1.1). The animals have a nervous system which sometimes features eyespots and other sensillae and a through gut with a complicated pharyngeal apparatus (see Chapter 4 for a discussion of internal anatomy).

Historical perspective

When the pioneering microscopists of the 18th Century first marvelled at the little *animalcules* they could see through their instruments, they were looking at animals that we now know as nematodes, rotifers and (the animals that are the subject of this book) tardigrades. The fascinating appearance of living tardigrades has encouraged observers to give them some very descriptive names such as Water Bear ('Wasser Bär') or Moss Piglet ('Mooschweinchen'). The first recorded observation of an animal described as a water bear is usually attributed to Goeze in 1773 (e.g. Ramazzotti and Maucci, 1982; Nelson, 1991) though there is some evidence that he was pre-empted by Eichhorn by six years (Owen, 1855). The current name, given to the animals in the 18th Century by Lazzaro Spallanzani, describes their lumbering gait (*tardi* - slow, *grado* - walker). The major events in the history of tardigrade research are summarized in Table 1.1.

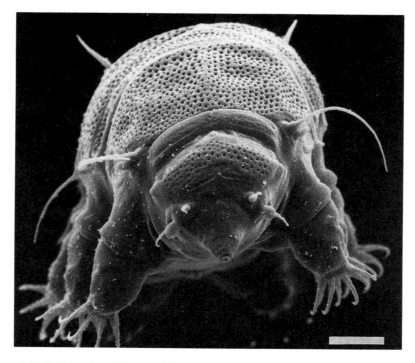

Figure 1.1. *En face* **view of a typical heterotardigrade 'Moss Piglet'**, *Echiniscus spiniger* SEM courtesy of Diane Nelson (bar ≈ 20 μm).

Table 1.1. Chronology of major events in tardigrade research

Year	Event
1674	Leeuwenhoek observes *animalcules* through his simple microscopes
1767	First claimed observation of a tardigrade by Eichhorn
1774	Corti notes the ability of tardigrades to revive after desiccation
1776	Spallanzani names the animal *Il Tardigrado*
1785	Müller corrects earlier observations of the number of legs to four pairs
1790	Linnaeus includes tardigrades in *Systema Naturae*
1834	Schultze describes *Macrobiotus hufelandi*. This species is named after Dr C.W. Hufeland who had earlier written a book on the art of prolonging human life
1838	Dujardin includes the Tardigrada in the group Systolidae with the Rotifera
1840	Schultze and Doyère independently describe the first heterotardigrade. Schultze's name *Echiniscus* is given priority over Doyère's *Emydium* as it was published three months earlier
1840	Doyère describes the genus *Milnesium* in the first monograph on tardigrades
1849	Boulengey, a student of Dujardin, observes the first marine tardigrade
1851	The first freshwater tardigrade (*Macrobiotus macronyx*) is described by Dujardin
1889	Tardigrades are considered to be primitive arthropods by Plate in the second monographic work to be published on the taxon
1900–1911	Numerous new species are described by the German, F. Richters
1905–1913	Numerous new species are described by the Scot, J. Murray
1928	Thulin revises the systematics of the taxon
1929	Marcus's monograph on tardigrades provides a bibliography of previously published work along with detailed descriptions of organ systems. The Tardigrada is considered a Class within the Arthropoda and split into the Heterotardigrada and Eutardigrada
1936	Marcus revises tardigrade systematics in his second monograph
1937	Rahm describes the third tardigrade class, the Mesotardigrada
1948	May publishes *La vie des tardigrades*
1962	Ramazzotti publishes the first edition of *Il Phylum Tardigrada* and recognizes the Tardigrada as a separate phylum
1964	The first discovery of a fossil tardigrade by Cooper
1968	Rosati publishes the first electron microscopic observations on tardigrades
1972	Ramazzotti publishes the second edition of *Il Phylum Tardigrada*
1974	First International Symposium on tardigrades in Pallanza, Italy
1977	Second International Symposium on tardigrades in Krakow, Poland
1980	Third International Symposium on tardigrades in Johnson City, U.S.A.
1980	Greven publishes *Die Bärtierchen*
1983	Ramazzotti and Maucci publish the third edition of *Il Phylum Tardigrada*
1985	Fourth International Symposium on tardigrades in Modena, Italy
1992	Fifth International Symposium on tardigrades in Maryland, U.S.A.
1994	Sixth International Symposium on tardigrades in Cambridge, U.K.

The slow rate of development in the knowledge of the tardigrades can, at least in part, be 'blamed' on the fact that tardigrades are of no economic, veterinary or medical importance. Parasitic species are rare and those that are known live on hosts that are also of little importance to Man's economy. The contrast with other invertebrate groups, such as insects and nematodes, could not be greater in this respect. The fact that after more than 200 years of microscopical investigation a whole phylum — the Loricifera (Kristensen, 1983) — can be discovered, has been described as 'emblematic' of how little we know of the microscopic living world (Wilson, 1992). Given that a whole phylum can be overlooked for so long, it seems probable that there is much to be discovered about the known phyla.

The fact that the tardigrade design or *Bauplan* is not as dominant as others (e.g. insects or crustaceans) does not make it any less significant, as the current number of extant species does not necessarily indicate the historical importance of a particular clade. Also the fact that tardigrades exhibit a range of extant species means that the tardigrade *Bauplan* is a success, in that the animals have been able to colonize geographically diverse habitats (terrestrial and marine) with little anatomic modification. In addition, the tardigrades may point the way to the ancestry of the great clade of tracheate arthropods, and this confers upon the group a much greater evolutionary importance than the number of extant species would suggest.

Aims of this book

The more accessible the literature on a zoological group, the more likely it is that students will be encouraged to study that group. This may explain the apparent resurgence of interest in tardigrades over the last twenty years or so, encouraged by the publication of the encyclopaedic monograph (in Italian) on the taxon by Ramazzotti and Maucci (1983); this publication draws together data sprinkled throughout the scientific literature (often in quite obscure journals) and provides a good foundation for further study. A programme of international symposia on the tardigrades has also been established during this time and has helped to bring together the members of the academic community interested in tardigrades. There have now been five international symposia dedicated to the Tardigrada, with a sixth in preparation at the time of writing. The proceedings of the first four have been published: Higgins (1975); Węglarska (1979*b*); Nelson (1982*b*) and Bertolani (1987*b*), providing an invaluable source of reference for the budding tardigradologist. During the twenty years since the first symposium, the number of described tardigrade genera has more than doubled (from 37 to 93). The increasing rate of description of new genera in the second half of the 20th Century, reaching a peak of 37 between 1981 and 1990 (Figure 1.2), is a result of a reorganization of the limno-terrestrial taxa (e.g. Kristensen, 1987; Pilato, 1987) and an increase in observations on marine taxa (particularly by Renaud-Mornant 1981; 1982*b*; 1983; 1984; 1989). It is hoped that this book will

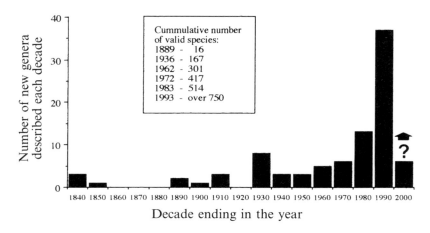

Figure 1.2. Graph showing the increase in the rate of generic descriptions during the second half of the 20th Century, with (inset) the number of species described in some of the most influential monographs

help to maintain the momentum generated by this group of enthusiasts by increasing the number of zoologists interested in tardigrades. The reader should not conclude from what has been said that the study of tardigrades is on the fringes of modern zoology — no text on general invertebrate zoology is complete without a discussion of the Tardigrada. However, an examination of a number of such texts will reveal that many rely exclusively on publications from the first half of this century as their sources, particularly the detailed drawings of tardigrade structure produced by Marcus (1929b; 1936). Two notable exceptions to this are the brief modern summaries given by Brusca and Brusca (1990) and Giere (1993). Therefore, it was felt that the time was right for a modern synthesis of tardigrade biology to highlight the advances made during the second half of this century, and particularly over the last two decades.

This book aims to give a balanced view of the current state of knowledge concerning the Tardigrada, and an extensive reference list is included to help guide those interested in pursuing more detailed investigations. However, in a book of this size, it is not possible (and probably not desirable) to refer to every article of interest concerning tardigrades — in particular, the vast numbers of publications describing new species and the occurrence of particular tardigrades in certain localities have not been listed unless they include other data pertinent to the text. In this respect, the bibliography is unbalanced as it under-represents those zoologists who have dedicated themselves to increasing our knowledge of the diversity of tardigrade species and providing zoogeographical data. Their work, however, should not be undervalued. A listing of the works concerned with the distribution of limno-terrestrial species has recently been prepared by McInnes (1994). A comprehensive tardigrade bibliography has also been prepared by Mackness (1994); although once the references have been identified, some will still prove difficult to locate.

In an attempt to avoid interrupting the flow of the text, the specialist terminology used is explained in the glossary towards the back of the book. It was felt important to include many of these terms as they will be encountered when examining the specialist literature cited in the text.

2

Origins and systematics

Fossil record

It is often valuable to study the fossil history when attempting to trace the ancestry of animal groups. Unfortunately, the fossil record of the Tardigrada consists of a single eutardigrade species (*Beorn leggi*) and a poorly preserved, unnamed juvenile heterotardigrade, both in Cretaceous amber (Cooper, 1964), from which it is impossible to infer much about the group as a whole except that there seems to have been very little change in tardigrade morphology over the past 60 million years. There is, however, a suggestion that some much older fossils from the Burgess Shales in Canada and from the Chengjiang deposits in China may give a clue to the origins of the group. These Cambrian animals (*Aysheaia pedunculata, Aysheaia prolata* and *Luolishania longicruris*) display similarities with the Annelida and the Onychophora, but also with the Tardigrada. These similarities and their phylogenetic significance have been widely discussed (Delle Cave and Simonetta, 1975; Simonetta, 1976; Renaud-Mornant, 1982a; McKenzie, 1983; Robison, 1985; Hou and Chen, 1989). The morphological characters which these worm-like animals have in common with tardigrades include the same type of poorly articulated limbs (lobopodia) terminating in claws, terminal mouths and the caudal end of the body merging into the last pair of legs. The circum-oral papillae of *Aysheaia* are very similar in morphology and arrangement to those of *Milnesium tardigradum*. However, the similarities between tardigrades and these ancient fossils are most pronounced when the primitive tardigrade genus *Parastygarctus* is used as the model (Renaud-Mornant, 1982a). Thus it is possible to hypothesize that an ancestral proto-tardigrade may have some traits in common with *Aysheaia* and *Parastygarctus*.

Valentine (1989) has suggested that arthropods and annelids share an ancestry among non-coelomate, segmented worms (from which they evolved independently), with the Tardigrada and Onychophora later diverging from the main arthropod lines. A model for this has been proposed by Dzik and Krumbiegel (1989) and is summarized in Figure 2.1. These authors also suggest

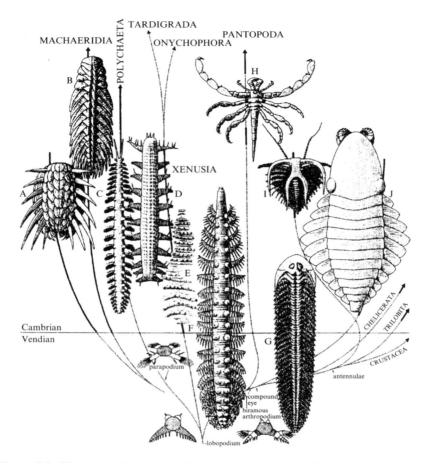

Figure 2.1. The most primitive fossil segmented animals and the proposed relationships of their classes
(A) *Wiwaxia*, (B) *Plumulites*, (C) *Burgessochaeta*, (D) *Aysheaia*, (E) *Plagiogmus*, (F) *Xenusion*, (G) *Spriggina*, (H) *Palaeoisopus*, (I) *Marrella* and (J) *Anomalocaris*. Not to scale. Points of origin of diagnostic features are indicated; diagrammatic cross-sections show the proposed original appearance of appendages in particular major groups. Reprinted from Dzik and Krumbiegel (1989), with permission.

placing the Tardigrada and Onychophora 'Classes' into the arthropodan subphylum Lobopodia, since they share so many characteristics which set them apart from more advanced arthropod groups.

Affinities of the Tardigrada

Since their discovery in the 18th Century, tardigrades have been placed in a number of taxonomic groups, sometimes alongside rotifers, annelids, insects, crustaceans, arachnids or mites (Ramazzotti and Maucci, 1982). There have been several recent attempts to allocate a position to the Tardigrada relative to other invertebrate groups (e.g. Inglis, 1985; Bergström, 1986; Willmer, 1990), with the tendency of aligning the Tardigrada with the Aschelminthes giving way to an alignment with the Arthropoda — particularly the Pararthropoda and with the Onychophora (Simonetta and Delle Cave, 1991; Kinchin, 1992a).

Detailed analyses of tardigrade structure have added much to the arguments of tardigrade phylogeny, but have not necessarily helped to reach an overall agreement, since authors examining the same structures have often arrived at conflicting conclusions, depending on the species observed and the techniques used. It must be stressed that many of the proposed relationships of the Tardigrada have only been tentative (see Table 2.1).

While the observations of these authors are not generally in dispute, the significance of their findings is. The question is whether or not the structures examined are plesiomorphic and, therefore, of phylogenetic importance, or whether they are simply examples of convergent evolution. On this point there seems to be little general agreement and doubts have been expressed about the validity of some proposed affinities.

- After examining the fine structure of tardigrade muscles and comparing his findings with observations of other invertebrate groups, Walz (1975a) concluded that the structure of muscle cells is independent of systematic position and phylogenetic relations and is determined exclusively by functional requirements.

Table 2.1. Tardigrade affinities as suggested by structural similarities

Feature analysed	Observed similarity	Reference
Viscerae	Aschelminthes	Clarke (1979)
Pharynx	Nematoda	Dewel and Clark (1973a)
Pharynx	Loricifera	Kristensen (1991)
Pharynx	Aschelminthes	Kinchin (1992a)
Buccal stylets	Nematoda	Riggin (1962)
Anhydrobiosis	Nematoda	Crowe and Madin (1974)
Claws	Pentastomida	Von Haffner (1977)
Claws	Annelida	Grimaldi de Zio et al. (1990b)
Claws	Onychophora	Ramsköld and Hou (1991)
Cuticle	Nematoda	Crowe et al. (1970)
Cuticle	Onychophora	Baccetti and Rosati (1971)
Cuticle	Arthropoda	Bussers and Jeuniaux (1973)
Sensillae	Arthropoda	Kristensen (1981)
Golgi beads	Arthropoda	Greven (1982)
Muscle attachments	Arthropoda	Shaw (1974)
Muscle cells	Arthropoda	Kristensen (1978)

- A tri-radiate lumen in the pharyngeal bulb has been observed in a number of taxa [see Kinchin (1992a) for a review] and has been cited as evidence of the aschelminth ancestry of the Tardigrada (e.g. Dewel and Clark, 1973). However, this has now been described as being of little phylogenetic importance as there is evidence that it has evolved a number of times (e.g. Ruppert, 1982) and simply represents the optimum way to pack and operate muscle around an extensible lumen (Bennet-Clark, 1976; Willmer, 1990).
- Lastly, the similar nature of the physiology governing desiccation tolerance (anhydrobiosis) in tardigrades and nematodes has also been cited as evidence of relatedness. However, a certain degree of convergence would be expected between animal groups that are subjected to common selection pressures in shared environments (Pilato, 1979). This is supported by the fact that anhydrobiosis is not exhibited by the primitive members of the phylum (certain marine heterotardigrades), but only by the more advanced terrestrial species. This is, therefore, a feature which has probably developed relatively late in the evolution of the tardigrades and so is separate from its development in the nematodes.

Bergström (1989) remains convinced that both the Aschelminthes and Tardigrada have a characteristic combination of features that is not a result of convergence, concluding that tardigrades arose from aschelminths by an additional arthropodization event. Kristensen (1991) also concludes that 'proarthropods' could have evolved from aschelminths rather than annelid-like animals. However, further data are required before this relationship can be established with any certainty.

Another enigmatic taxon that is often described as having uncertain affinities and grouped with the Tardigrada, is the Pentastomida[1]. While their parasitic life-style has led to some anatomical simplifications, these animals (particularly the larvae) exhibit a number of striking structural similarities with tardigrades: a sac-like gut (the anterior end of which pumps liquid); a lack of gas exchange and circulatory organs; legs with claws; a caudal extension (cerci in pentastomids/hind legs in tardigrades); and sensory papillae (Figure 2.2). Von Haffner (1977) considers the pentastomids to lie between tardigrades and myriapods.

Valentine (1989) cites evidence that arthropodization occurred a number of times to produce the Uniramia, Crustacea, Chelicerata, Trilobita and other groups, possibly including the Tardigrada. The argument about the Tardigrada being a group within the Arthropoda, or a group which has converged towards an arthropodan habit, will certainly not be concluded in this book or in the near future. However, given the arguments for multiple arthropodization events

---

[1] *The Pentastomida were previously considered to be a separate phylum, but are now more generally considered to be arthropods, see Storch (1993) for a review.*

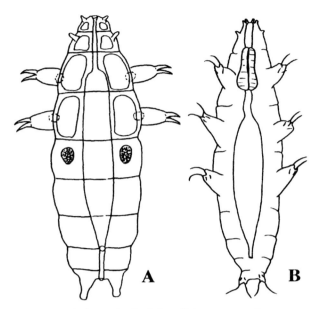

Figure 2.2. Comparison of a hypothetical, free-living ancestral pentastomid larva (A), redrawn from Von Haffner (1977), and the eutardigrade *Milnesium tardigradum* (B)

(e.g. Manton, 1977; Valentine, 1989; Willmer, 1990), and the ultrastructural evidence that tardigrades are arthropodan (e.g. Baccetti and Rosati, 1971; Kristensen, 1978), a possible polyphyletic arthropod scheme giving a position to the Tardigrada within the Uniramia is suggested (Figure 2.3). The number and position of the major arthropodization events are debatable.

Recent observations of *Kerygmachela kierkegaardi* from the Sirius Passet deposits in Greenland (Budd, 1993) have added a further complication by suggesting that the ancient lobopods are more closely related to the biramous arthropods (crustaceans) than to the Uniramia. This finding does, however, add weight to the arguments for a polyphyletic (at least biphyletic) origin of the arthropod clade. It is also possible that different lobopod groups gave rise to biramous and uniramous arthropods independently (Dzik and Krumbiegel, 1989).

Arthropod polyphyly does not necessarily imply independent evolution of all arthropod characters. It has been shown that the main locomotory and feeding mechanisms of the arthropod subgroups were probably derived independently from different soft-bodied ancestors (Manton, 1977). These may well have had common ancestors themselves, but the fact that they were not arthropodized means that they cannot be considered as arthropods. Therefore, the later 'arthropod grade' is polyphyletic. It can be seen from this that the key point in these arguments is the boundary, arbitrarily set between the ancestral and descendant groups. Groups which are polyphyletic in one sense can always be 'made' monophyletic by adjusting the boundary and choosing more widely distributed diagnostic characters (Cloudsley-Thompson,

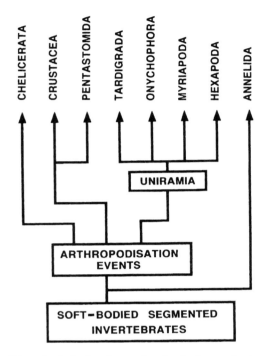

Figure 2.3. A possible polyphyletic scheme for the Arthropoda showing the relative position of the Tardigrada with the other groups within the Uniramia

1988). In consequence, views on arthropod phylogeny are as divergent as the animals in question.

This leaves the classic annelid–uniramian evolutionary line from which tardigrades are thought to have emerged. The Onychophora are often placed as intermediary forms in an evolutionary progression from the annelids to the arthropods (e.g. Strickberger, 1990) with the Tardigrada branching off at some ill-defined point along the line (Figure 2.4). This is supported by ultrastructural studies which have been able to place the Tardigrada within the annelid–uniramian evolutionary line (e.g. Baccetti and Rosati, 1969; Baccetti et al. 1971; Węglarska 1979a).

Systematics of the phylum

The separation of the Tardigrada into a distinct phylum by Ramazzotti (1962) has been described by Greven (1982) as a temporary solution which only sidesteps the issue of the group's phylogeny. However, none of the schemes which have allocated a systematic position to the Tardigrada has been universally accepted and the maintenance of the phylum status for the Tardigrada is the system which is likely to cause least confusion, though given the scheme proposed in Figure 2.3, it may make more sense to describe the Tardigrada as a subphylum within the Phylum Uniramia along with the Onychophora,

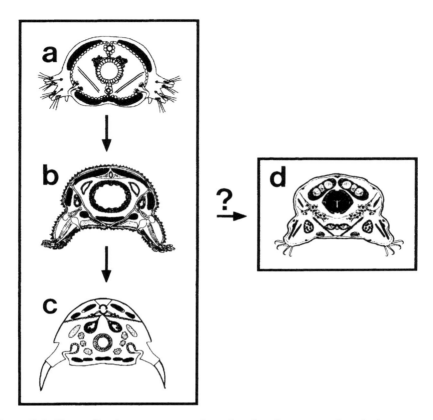

Figure 2.4. Generalized transverse sections showing the proposed evolutionary progression from annelid parapodes (a) through onychophoran lobopodes (b) to the true articulated limbs of the arthropods (c); the Tardigrada (d) are thought to have branched off from this line

Hexapoda and Myriapoda. While other views will doubtless be put forward in the future, the Tardigrada is recognized as a phylum throughout this work in order to retain the sense of the vast majority of modern taxonomic tardigrade papers.

There have been a number of major systematic works on tardigrades published this century. Of these, the most influential have probably been those by Thulin (1928), by Marcus (1929*b*; 1936) — whose morphological observations have only been surpassed by later workers who were able to use electron microscopes to elucidate greater detail — and the three editions of *Il Phylum Tardigrada* by Ramazzotti (1962; 1972) and then Ramazzotti and Maucci (1983), which combined observation of the structure and habits of tardigrades with detailed descriptions of all the species known to date. These works formed the basis of modern tardigrade taxonomy.

The major taxa within the Tardigrada were established by Marcus (1929*b*) who split the group in two, forming the Heterotardigrada and Eutardigrada (see Table 2.2). These are now considered as classes within the phylum and are

Table 2.2. Tardigrade classification

Marine heterotardigrades

Order	Family	Subfamily	Genus
Arthrotardigrada	Halechiniscidae	Orzeliscinae	*Orzeliscus* Du Bois-Reymond Marcus, 1952[*]
			Opydorscus Renaud-Mornant, 1989
		Halechiniscinae	*Halechiniscus* Richters, 1908[*]
			Chrysoarctus Renaud-Mornant, 1984
			Paradoxipus Kristensen and Higgins, 1989
		Florarctinae	*Florarctus* Delamare-Deboutteville and Renaud-Mornant, 1965
			Wingstrandarctus Kristensen, 1984
			Ligiarctus Renaud-Mornant, 1982b
		Styraconyxinae	*Styraconyx* Thulin, 1942
			Angursa Pollock, 1979
			Lepoarctus Kristensen and Renaud-Mornant, 1983
			Pleocola Cantacuzéne, 1951
			Raiarctus Renaud-Mornant, 1981
			Rhomboarctus Renaud-Mornant 1984
			Tetrakentron Cuénot, 1893
			Bathyechiniscus Steiner, 1926
			Tholoarctus Kristensen and Renaud-Mornant, 1983
			Paratanarctus D'Addabbo Gallo et al. 1992
		Euclavarctinae	*Euclavarctus* Renaud-Mornant, 1975
			Clavarctus Renaud-Mornant, 1983
			Exoclavarctus Renaud-Mornant, 1983
			Proclavarctus Renaud-Mornant, 1983
			Parmursa Renaud-Mornant, 1984
			Moebjergarctus Bussau, 1992
		Tanarctinae	*Tanarctus* Renaud-Debyser, 1959
			Actinarctus Schulz, 1935

Table 2.2. contd.

Marine heterotardigrades

Order	Family	Subfamily	Genus
	Stygarctidae	Archechiniscinae	Zioella Renaud-Mornant, 1987
			Archechiniscus Schulz, 1953
		Stygarctinae	Stygarctus Schulz, 1951
			Pseudostygarctus McKirdy et al., 1976
			Mesostygarctus Renaud-Mornant, 1979
			Parastygarctus Renaud-Debyser, 1965
			Megastygarctides McKirdy et al., 1976*
		Neoarctinae	Neoarctus Grimaldi De Zio et al., 1992
		Neostygarctinae	Neostygarctus Grimaldi de Zio et al., 1982
	Renaudarctidae		Renaudarctus Kristensen and Higgins, 1984a
	Coronarctidae		Coronarctus Renaud-Mornant, 1974
	Batillipedidae		Batillipes Richters, 1909*
Echiniscoidea	Echiniscoididae		Anisonyches Pollock, 1975c
			Echiniscoides Plate, 1889*

Marine Eutardigrades

Order	Family	Subfamily	Genus
Parachela	Hypsibiidae	Hypsibiinae	Halobiotus Crisp and Kristensen, 1983

Limno-terrestrial heterotardigrades

Order	Family		Genus
Echiniscoidea	Oreellidae		Oreella Murray, 1910
			Carphania Binda, 1978
	Echiniscidae		Echiniscus Schultze, 1840
			Antechiniscus Kristensen, 1987
			Bryochoerus Marcus, 1936
			Bryodelphax Thulin, 1928
			Cornechiniscus Maucci and Ramazzotti, 1981

Table 2.2. contd.

Limno-terrestrial heterotardigrades

Order	Family	Genus
Echiniscoidea (contd)	Echiniscidae	Hypechiniscus Thulin, 1928
		Mopsechiniscus du Bois-Reymond Marcus, 1944
		Novechiniscus Kristensen, 1987
		Parechiniscus Cuénot, 1926
		Proechiniscus Kristensen, 1987
		Pseudechiniscus Thulin, 1928
		Testechiniscus Kristensen, 1987

Limno-terrestrial eutardigrades

Order	Family	Subfamily	Genus
Parachela	Macrobiotidae		Macrobiotus Schultze, 1834*
			Murrayon Bertolani and Pilato, 1988*
			Minibiotus Schuster et al., 1980*
			Macroversum Pilato and Catanzaro, 1988
			Pseudodiphascon Ramazzotti, 1964
			Dactylobiotus Schuster et al., 1980*
			Adorybiotus Maucci and Ramazzotti, 1981
			Richtersius Pilato and Binda, 1987b; 1989*
			Calcarobiotus Dastych, 1993
	Eohypsibiidae		Eohypsibius Kristensen, 1982
			Amphibolus Bertolani, 1981b
	Calohypsibiidae		Calohypsibius Thulin, 1928*
			Haplomacrobiotus May, 1948
			Hexapodibius Pilato, 1969
			Parhexapodibius Pilato 1969
			Haplohexapodibius Pilato and Beasley, 1987
	Necopinatidae		Necopinatum Pilato, 1971

Table 2.2. contd.

Limno-terrestrial eutardigrades

Order	Family	Subfamily	Genus
	Hypsibiidae	Hypsibiinae	*Hypsibius* Ehrenberg, 1848[*]
			Microhypsibius Thulin, 1929[*]
			Ramazzottius Binda and Pilato, 1986[*]
			Eremobiotus Biserov, 1992
			Doryphoribius Pilato, 1969
			Isohypsibius Thulin, 1928[*]
			Mixibius Pilato, 1992
			Pseudobiotus Schuster et al., 1980[*]
			Ramajendas Pilato and Binda, 1990
			Thulinia Bertolani, 1981b
		Itaquasconinae	*Itaquascon* De Barros, 1939
			Mesocrista Pilato, 1987[*]
			Platicrista Pilato, 1987[*]
			Parascon Pilato and Binda, 1987a
		Diphasconinae	*Diphascon* Plate, 1889[*]
			Paradiphascon Dastych, 1992
			Hebesuncus Pilato, 1987[*]
			Fujiscon Ito, 1991
Apochela	Milnesiidae		*Milnesium* Doyère, 1840[*]
			Limmenius Horning et al., 1978
Incertae sedis			*Apodibius* Dastych, 1983
			92 genera

[*]Only those marked have been observed from the British Isles.

Details of genera described before 1983 can be found in Ramazzotti and Maucci (1983), protologues of those described more recently are listed in the references.

sometimes described colloquially as 'armoured' and 'naked' forms, respectively (Figure 2.5). This refers to the cuticular dorsal plates found in terrestrial heterotardigrades, but absent in eutardigrades.

A third class of tardigrade, the Mesotardigrada, was established on the basis of the description of *Thermozodium esakii* (Figure 2.6a) by Rahm (1937). This species was discovered in a hot spring near Nagasaki, Japan. However, since neither type material nor type locality has survived to the present day, and no other species of mesotardigrade has yet been described, this poorly documented group will not be discussed any further here.

Another strange tardigrade whose systematic position has caused problems was described by Iharos (1968). A single specimen of *Echinursellus longiunguis* from Chile (Figure 2.6b) was originally considered to be a freshwater missing link between the two heterotardigrade orders, Echiniscoidea and Arthrotardigrada. However, after re-examination of the material by Kristensen (1987), this specimen is now considered to be a eutardigrade of the genus *Pseudobiotus*.

While there is a recognition of the roles of advanced techniques such as karyological analysis (Bertolani, 1975; 1976; 1982a; Bertolani *et al.* 1987; Redi and Garagna, 1987; Rebecchi, 1991) and molecular systematics (Berlocher, 1982), tardigrade systematics is at present firmly rooted in morphological comparisons. The main characters used in heterotardigrade systematics are cephalic appendages, cuticular extensions and claws (particularly in the Arthrotardigrada), and also the pattern of the dorsal cuticular plates (in the terrestrial Echiniscoidea) (see Chapter 3). In the past, marine tardigrades (most of which belong to the Arthrotardigrada) have been given less attention than their terrestrial relatives. Marcus (1936) listed 274 species, of which only six were marine, and consequently the systematics of marine species is much more fluid than that of the limno-terrestrial species. Almost nothing is known about intraspecific variation for most marine species because of the difficulties in sampling. The wide morphological differences exhibited by marine forms have resulted in the creation of numerous genera (Table 2.2), many of which are monotypic. Only in recent years have some of the more commonly observed marine tardigrades (e.g. *Echiniscoides* spp.) been examined closely enough and in large enough numbers to separate species (e.g. Kristensen and Hallas, 1980; Bellido and Bertrand, 1981; Hallas and Kristensen, 1982; D'Addabbo Gallo *et al.*, 1992).

The major morphological characteristics which are used in eutardigrade systematics are the claws and the buccopharyngeal apparatus (see Chapters 3 and 4). The Eutardigrada tends to be a more homogeneous group morphologically than the Heterotardigrada, resulting in fewer genera, each containing more species. During the 1980s, as more genera were being described, conflicting views of their arrangement within the eutardigrade families were emerging. Some systems placed more emphasis on the structure of the buccopharyngeal apparatus (e.g. Schuster *et al.*, 1980) and others concentrated more on claw structure (e.g. Pilato, 1982). After detailed examination of intraspecific variation,

Origins and systematics

Figure 2.5. Antero-lateral views of a typical limno-terrestrial heterotadigrade, *Echiniscus mauccii* (a) and a typical eutardigrade, *Macrobiotus tonollii* (b)
SEMs courtesy of Diane Nelson (bars ≈ 20 μm).

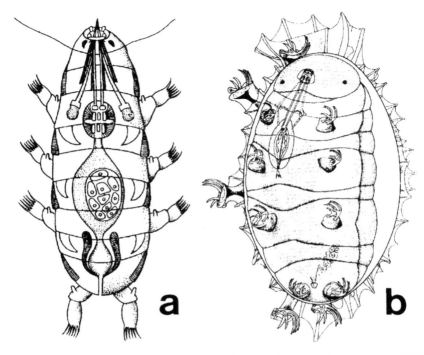

Figure 2.6. (a) *Thermozodium esakii* (Mesotardigrada) [reprinted from Rahm (1937), with permission from the Zoological Society of Japan] and (b) *Echinursellus* (now *Pseudobiotus*) *longiunguis* [reprinted from Kristensen (1987), with permission from Mucchi Editore, Italy]

claw structure is now considered a more conservative character than buccopharyngeal apparatus and is, therefore, the fundamental character for distinguishing different families of eutardigrades (Bertolani and Kristensen, 1987). The sequence of claw branches with respect to the midline of extended legs is, in some genera, alternate (2-1-2-1) (i.e. secondary-primary-secondary-primary) as in *Calohypsibius*. In other genera, the two primary branches are adjacent to each other (2-1-1-2), as in *Dactylobiotus*. However, members of an unusual genus described by Dastych (1983), *Apodibius*, have reduced legs totally lacking claws. The absence of this important character has resulted in the unsatisfactory consideration of the genus as *incertae sedis* (Table 2.2). The most likely position for this genus is probably in the Family Calohypsibiidae alongside *Parhexapodibius* and *Hexapodibius*, which display a lesser reduction in legs and claws and may illustrate an evolutionary trend (Pilato, 1989).

The systematic arrangement given in Table 2.2 is a synthesis of the work of numerous authors, most notably Pilato (1969; 1972; 1975; 1982; 1987); Schuster *et al.* (1980); Ramazzotti and Maucci (1983); Renaud-Mornant (1984); Kristensen (1987) and Dastych (1992). With the exception of the genus *Halobiotus* and two *Isohypsibius* spp. (Hallas, 1971; Tsurusaki, 1980; Kristensen, 1982*a*; Crisp and Kristensen, 1983), the eutardigrade genera are generally terrestrial types. However, on rare occasions a few species which are regarded as terrestrial have been recovered from marine habitats (e.g. Renaud-Debyser, 1964; Morgan, 1976) and one arthrotardigrade, *Styraconyx hallasi*, has been observed in freshwater (Kristensen, 1977). The marine/freshwater divide used in Table 2.2 should, therefore, be observed with caution.

It is clear that the systematics of the Phylum will go through a number of changes as the holes in our knowledge are slowly plugged. The positions allocated to genera such as *Apodibius* and *Hebesuncus* [whose position within the subfamily Diphasconinae been described as 'tentative' by Dastych (1992)], are likely to be the subject of debate. It is also likely that as new criteria are developed for the description of genera, more will be erected. These will probably include new species, but existing species may well be re-assigned as has previously been the case when new genera have been described, for example, *Amphibolus* (Bertolani, 1981*a*) and *Ramazzottius* (Binda and Pilato, 1986).

As a result of the large morphological diversity exhibited by the known genera of marine heterotardigrades, those constructing phylogenetic models have tended to concentrate on this class. The marine heterotardigrades are generally considered to be the most primitive members of the phylum (Renaud-Mornant, 1982*a*; Kristensen and Renaud-Mornant, 1983; Grimaldi de Zio *et al.* 1987) with the Neostygarctidae and Renaudarctidae as the most primitive of the heterotardigrades (Grimaldi de Zio, 1986). Two evolutionary lines have been proposed to stem from the Neostygarctidae: the first leading towards the Batillipedidae and the second leading towards the terrestrial Echiniscidae (and possibly from there on to the Eutardigrada). Evidence for these evolutionary lines comes largely from observations of the structure of the claws (Kristensen and Higgins, 1984*a*; Grimaldi de Zio *et al.*, 1987) and is summarized in Figure 2.7. In the first line, toes with claws give way to toes with adhesive discs. In the second line, the toes disappear altogether so that the claws are inserted directly on to the tarsus — the condition found in the Echiniscoidea and Eutardigrada. Vestigial toes are possibly represented by the papillae on the legs of the Echiniscidae (Schulz, 1953).

This second line has been followed through the Echiniscidae at the generic level by Kristensen (1987), although he believes the Echiniscidae to be more closely related to the Stygarctidae and Renaudarctidae than was proposed by Grimaldi de Zio *et al.* (1987) because they have similar dorsal plates. Classically, the dorsal plates have been used to show the different lines within the Echiniscidae (Figure 2.8). The two main lines are the *Echiniscus* line and the *Pseudechiniscus* line — separated by the presence or absence of a pseudosegmental plate on the dorsum (see Chapter 3). Kristensen (1987) concludes that the heterotardigrades must have invaded terrestrial habitats very early in the Palaeozoic Era, although fossil evidence older than the Cretaceous Period (Cooper, 1964) has not been found. He also concludes that as all echiniscids are herbivorous (feeding on bacteria, algae, bryophytes and lichens), their evolution probably followed that of the bryophytes (see Chapter 7).

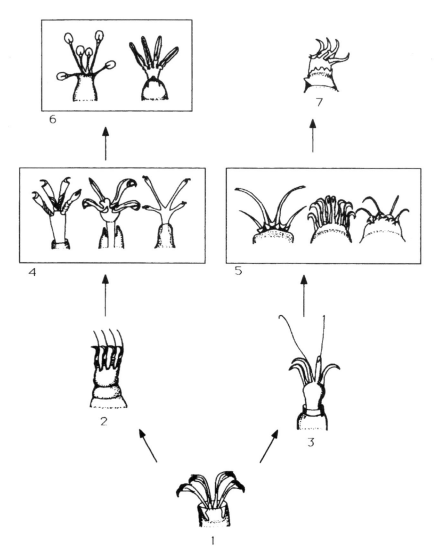

Figure 2.7. Proposed phylogenetic relationships among the Heterotardigrada based on claw structure
(1) Neostygarctidae, (2) Renaudarctidae, (3) Stygarctidae (4) Halechiniscidae, (5) Coronarctidae and Echiniscoididae, (6) Batillipedidae and Orzeliscidae and (7) Echiniscidae. Redrawn from Grimaldi de Zio et al. (1987).

Evolutionary relationships in the Eutardigrada have been re-defined by Pilato (1969) on the basis of the location, structure and shape of the claws and the structure of the buccal apparatus. Four main evolutionary lines were proposed, corresponding to the four families Macrobiotidae, Hypsibiidae, Calohypsibiidae and Milnesiidae (see Table 2.2). Two other such lines were added later, the Necopinatidae (Ramazzotti and Maucci, 1983) and the

Origins and systematics

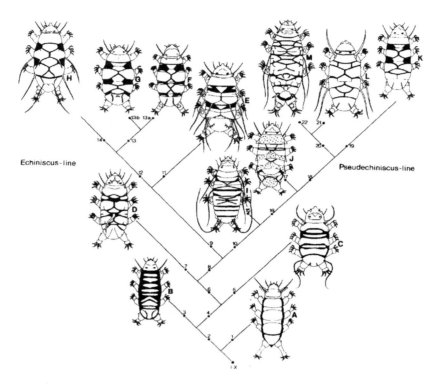

Figure 2.8. Cladogram of habitus drawings (females) of the 12 genera in the Echiniscidae and the sister group Oreella
(A) *Oreella mollis*, (B) *Parechiniscus chitonides*, (C) *Novechiniscus armadilloides*, (D) *Hypechiniscus gladiator*, (E) *Testechiniscus spitsbergensis*, (F) *Bryochoerus intermedius*, (G) *Bryodelphax sinensis*, (H), *Echiniscus merokensis*, (I) *Proechiniscus hanneae*, (J) *Antechiniscus lateromamillatus*, (K) *Pseudechiniscus suillus*, (L) *Mopsechiniscus imberbis* and (M) *Cornechiniscus holmeni*. Reproduced from Kristensen (1987), with permission from Mucchi Editore, Italy.

Eohypsibiidae (Bertolani and Kristensen, 1987). The Macrobiotidae consists of those forms in which the claws of each limb are symmetrical with respect to the median plane of the limb, generally exhibiting a 2-1-1-2 sequence. In the Hypsibiidae, the claws are asymetrical and of different size and shape. The Hypsibiidae is further divided on the basis of the buccal apparatus (in the Hypsibiinae the buccal tube is rigid, in the Itaquasconinae and Diphasconinae the buccal tube has a flexible portion — see Chapter 4). The *Calohypsibius*-type claws are similar in size and shape exhibiting an alternate (2-1-2-1) sequence. The claws of the Eohypsibiidae each have three distinct regions, while those of the Necopinatidae are small, unbranched structures which are only present on the first pair of legs. The Milnesiidae is separated from the other families on the basis of the morphology of its sensillae, in addition to claw and buccal apparatus structure (see Chapter 9, *Milnesium tardigradum*). Details of the structure of claws and buccal apparatus are given in Chapters 3 and 4.

3

External morphology

Morphological groups

The following summary of tardigrade external morphology is divided into three sections, each discussing one of the three main morphological groups.

- The arthrotardigrades are marine forms which have well-developed cephalic appendages, usually including a median cirrus and always with cirrus A. The extremities of the legs are often, but not always, digitate.
- The echiniscoids are limno-terrestrial armoured forms and marine non-armoured forms. The cephalic appendages are less well-developed than in arthrotardigrades, the median cirrus is usually absent, but cirrus A is always present. The extremities of the legs are non-digitate.
- The eutardigrades are mostly limno-terrestrial forms. The cephalic appendages (including the median cirrus and cirrus A) are absent. The cuticle is not armoured. The extremities of the legs are non-digitate. The double claws are differentiated into a primary and secondary arm.

Arthrotardigrada

The cephalic sensory appendages are among the most useful taxonomic characters for the Arthrotardigrada (Figure 3.1) and a full set comprises a single median cirrus; one pair of internal (buccal) cirri; one pair of external (buccal) cirri; one pair of cirri A (lateral cirri); one pair of primary clava; one pair of secondary clava (cephalic papillae); and one pair of tertiary clava (rare).

In a number of genera, such as *Parastygarctus*, these appendages are borne on cuticular projections of diverse size and shape which may drastically alter the appearance of the animal and may cause confusion to the observer. However, the relative positions of the appendages remain more or less unaltered. The cirri are of variable size. Each is composed of three substructures: a bulbous base (cirrophore), a collar (scapus) and a hair-like flagellum (Figure 3.1).

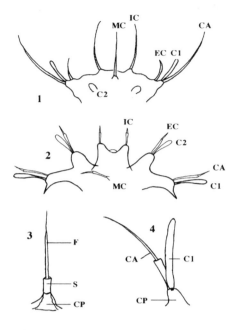

Figure 3.1. Cephalic sense organs of the Arthrotardigrada
(1) *Batillipes*, (2) *Parastygarctus*, (3) cirrus structure and (4) cirrus A and primary clavus on a common cirrophore. Abbreviations used: MC, median cirrus; CA, cirrus A (lateral cirrus); IC, internal buccal cirrus; EC, external buccal cirrus; C1, primary clavus; C2, secondary clavus; F, flagellum; S, scapus; and CP, cirrophore.

The relative sizes of these three components are used as taxonomic criteria within this group. A refractive structure, a small tube, has been observed in the cirrophore of the primary clavus of a few species and after its original observation in *Florarctus antillensis* by Van der Land (1968) was subsequently called 'Van der Land's organ by Kristensen and Higgins (1984*b*). This term is now widely used in the description of marine tardigrades (e.g. Grimaldi de Zio *et al.*, 1990*a*). Tertiary clava are rare and have only been observed in very few species in the genera *Angursa*, *Renaudarctus* and *Paradoxipus*. Clava display variable morphology, with examples being club-shaped, dome-shaped, banana-shaped or kidney-shaped. The seminal receptacles of some species are quite noticeable and are of variable shape and size (Figure 3.2). Noda (1986) has suggested that the shape of the vesicle and duct forming the seminal receptacles of the Halechiniscidae can also be a useful taxonomic character at the subfamily level (Table 3.1). However, some species do not appear to conform to Noda's scheme, e.g. the ducts of the seminal receptacles in *Styraconyx tyrrhenus* are sinuate rather than coiled (D'Addabbo Gallo *et al.*, 1989) suggesting that some revision of Noda's scheme is necessary before it may be used with any confidence.

Van der Land (1968) described the legs of *Florarctus* as 'arthropoda', in that they are true jointed limbs, but made no conclusion about the phylogenetic significance of this condition. The terminology adopted is, however, that used for the eu-arthropod limb (Figure 3.3). The variety of claws exhibited by the Arthrotardigrada are illustrated in Figure 2.7. A number of species exhibit extreme development of the lateral filaments and/or lateral or caudal extensions of the cuticle (alae), examples of which are illustrated in Figure 3.4. These are probably an aid to dispersal in the marine environment (see Chapter 7).

Figure 3.2. Ventral view of the caudal region of *Neostygarctus acanthophorus* showing the gonopore (G), anus (A), vesicle (V) and duct (D) of the seminal receptacles
Redrawn from Grimaldi de Zio et al. (1990b).

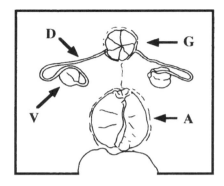

Table 3.1. Taxonomic trends in the morphology of seminal vesicles of halechiniscids

	Vesicle	Duct
Florarctus-type	Moderately large	Medially thickened
Tanarctus-type	Indistinct	Thick
Euclavarctus-type	Elongate	Long and sinuate
Styraconyx-type	± Spherical	Coiled
Halechiniscus-type	Ovoid	Short and s-shaped

After Noda, 1986.

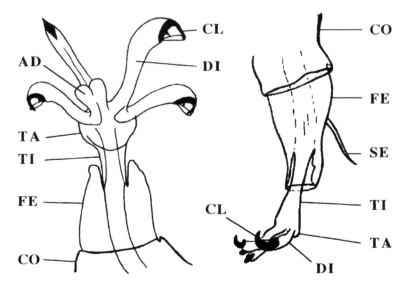

Figure 3.3. Typical halechiniscid claw and leg morphology
Redrawn and simplified from Kristensen (1977) and Van der Land (1968). Abbreviations used: AD, adhesive pad; CL, claw; DI, digit; TA, tarsus; TI, tibia; FE, femur; CO, coxa; and SE, seta.

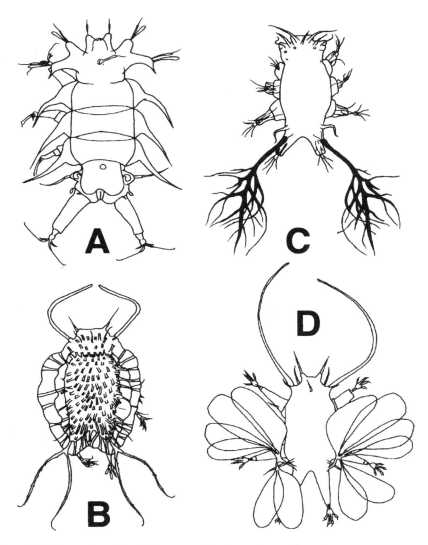

Figure 3.4. Examples of cuticular extensions in arthrotardigrades
(A) *Parastygarctus sterreri*, (B) *Actinarctus lyrophorus*, (C) *Tanarctus dendriticus* and (D) *Tanarctus velatus*. Redrawn from various authors.

Echiniscoidea

In the limno-terrestrial echiniscids, the plates on the dorsal cuticle are a distinctive feature (Figure 3.5) and one of the most important in determining species. These plates are absent in marine forms such as *Echiniscoides* spp. (see Chapter 9). The texture of the plates is highly variable (Figure 3.6). In a given species, the texture may be even across the plates or may vary from plate to plate with for example, the texture of the scapular plate being very different to that of the terminal plate. Alternatively, plates may be subdivided by bands of varying

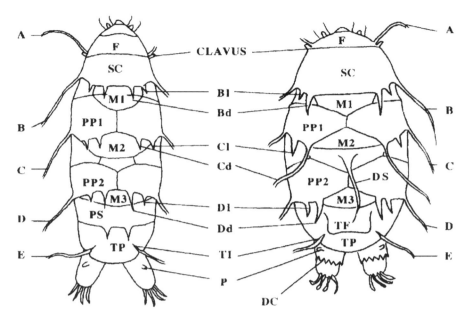

Figure 3.5. Cuticular terminology for echiniscids, *Pseudechinscus*-type (left) and *Echiniscus*-type (right)
Abbreviations used: A–E, Cirri A–E; F, frontal plate; SC, scapular plate; M1–3, median plates 1–3; PP1 and 2, paired plates 1 and 2; PS, pseudosegmental plate; TP, terminal plate; TF, terminal facet; TI, terminal indentation; P, papillus; DC, dentate collar; MC, median cirrus; Bl, Cl and Dl, lateral spines B, C and D; and Bd, Cd and Dd, dorsal spines B, C and D. Spines can also occur as filaments (as shown for Cd in *Echiniscus*). M1–3 can be divided (as shown here in M2). Dorsal cirrus only occurs in *Hypechiniscus* sp. Cirrus A is always present, but all other cirri and spines may be present or absent in any combination. Spines can also be present on the posterior margin of the pseudosegmental plate.

texture as in *Echiniscus trisetosus* (Figure 3.7A). The appearance of the texture may vary with the plane of focus when viewed under high magnifications. It should, therefore, be viewed in various planes. A number of species descriptions have included detail of the cuticle when viewed with the objective in both the 'high' and 'low' focus positions. However, the plane of focus is difficult to standardize and has added confusion to some descriptions.

Pigmentation is variable, with some species very brightly coloured (Figure 3.8). The pigmentation of some species is thought to be derived from the animal's food source, e.g. bright orange from the carotenoids available from many lichens (Massonneau and May, 1950). The movement of carotenoids from the gut lumen to the cuticle via the body cavity has been described by Mihelčič (1950). The shape of the plates is also highly variable. Typical forms of echiniscid genera are shown in Figure 2.8. In addition to variations in their size and texture, the plates may be fused or subdivided, for example, in *Novechiniscus* spp. all the segmental plates are unpaired. The single plates span the width of the dorsum. In *Testechiniscus* spp. ventral plates (sternites) are present in addition to those on the dorsum. In *Echiniscus tesselatus* the scapular

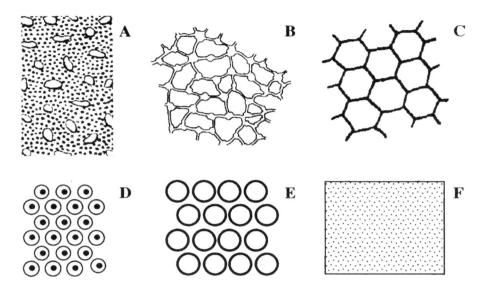

Figure 3.6. Examples of cuticle texture in the Echiniscidae
(A) *Echiniscus quadrispinosus*, (B) *Echiniscus rackae*, (C) *Echiniscus tardus*, (D) *Echiniscus rosaliae*, (E) *Echiniscus ramazzottii* and (F) fine granulation common in many species.

plate is divided into 10 platelets (Figure 3.7B). In many species the terminal plate exhibits two indentations which may be only small notches in the posterior margin of the plate or may extend across to the anterior margin of the plate, effectively dividing it into platelets, as in *Hypechiniscus gladiator*. In a number of species such as *Pseudechiniscus novazeelandiae* the posterior margin of the pseudosegmental plate bears appendages (Figure 3.7C).

Within some echiniscid species, there is considerable variation of the spines and filaments originating from the cuticle. For example, in *Hypechiniscus gladiator* a number of different forms and subspecies have been described exhibiting obvious differences in the size and number of dorsal spines (Figure 3.9). In addition to those described by Iharos (1973), Morgan and King (1976) decribe a form of *Hypechiniscus gladiator* from Ben Nevis in Scotland with five dorsal spines. It seems likely that other variations will be found in the future.

Another species which is regarded as highly variable is *Echiniscus merokensis* (Figure 3.10). Ramazzotti and Maucci (1983) affirm that the forms with spine Bd on the scapular plate certainly are *Echiniscus merokensis suecica*, but when examined outside mixed populations, 'no tardigradologist would recognize it as such'.

In a number of species of heterotardigrade, the mouth is positioned at the end of a retractable mouth cone (Figure 1.1). The typical rosette-like appearance of the gonopore is shown in Figures 3.2 and 5.1. Four single claws are inserted directly on the leg. On the fourth pair of legs in *Echiniscus* spp., a dentate collar is visible (Figures 3.8 and 3.11). This has a variable number of teeth. Papillae are also often visible on the leg (Figure 3.11).

External morphology

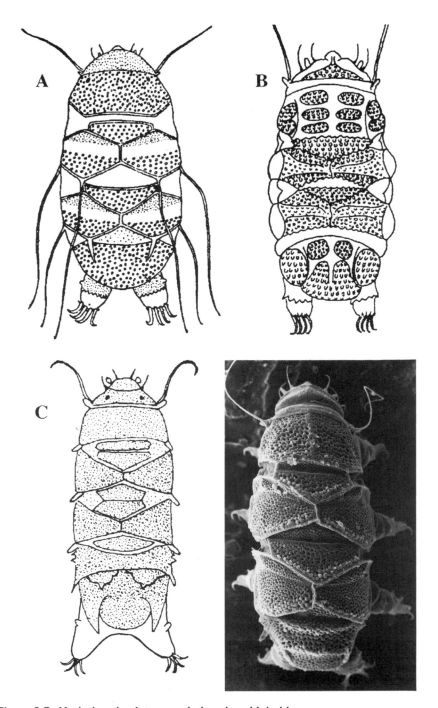

Figure 3.7. Variations in plate morphology in echiniscids
(A) *Echiniscus trisetosus*, (B) *Echiniscus tesselatus* and (C) *Pseudechiniscus novazeelandiae* forma *marinae*. SEM shows *Echiniscus mauccii*. A–C reproduced from Ramazzotti and Maucci (1983), with permission. SEM courtesy of Diane Nelson.

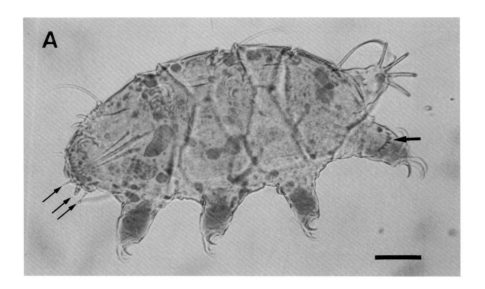

Figure 3.8. Photographs of a fresh specimen of *Echiniscus granulatus* showing natural coloration
(A) Lateral view: the cephalic papilla with buccal cirri on either side are visible at the anterior end (three arrows), while the dentate collar is visble on the fourth pair of legs (single arrow). (B) Dorsal view showing the arrangement of dorsal plates and the cuticular sculpture. Bars = 50 μm.

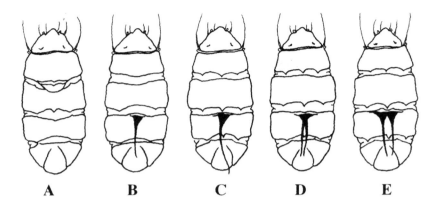

Figure 3.9. Varieties of *Hypechiniscus gladiator*
(A) Forma *exarmata*, (B) forma *nominal*, (C) forma *spinulosa*, (D) forma *fissigladii* and (E) forma *bigladii*. Redrawn from Iharos (1973).

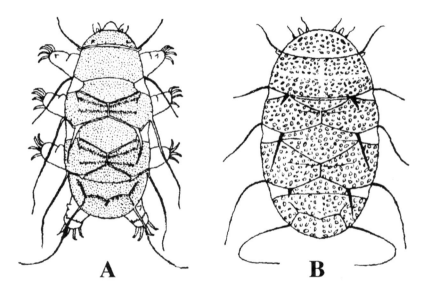

Figure 3.10. *Echiniscus merokensis* forma *suecica*
(A) Typical form, (B) rare form with spines on the scapular plate. Reproduced from Ramazzotti and Maucci (1983), with permission.

Eutardigrada

Of the external structures visible in eutardigrades, the claws are the most important in identifying taxa. The basic claw structure and the terminology used to describe it is shown in Figure 3.12. Eutardigrade claws are highly variable in shape (Figure 3.13). In the order Parachela, the secondary branch is attached to the leg and the primary branch arises from the secondary branch. In the Macrobiotidae, the two double claws on each leg are similar in size and

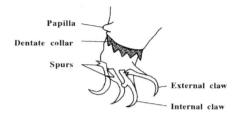

Figure 3.11. Detail of echiniscid claw

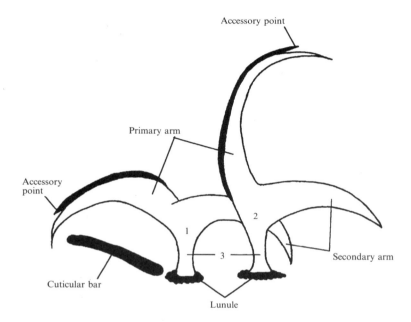

Figure 3.12. Eutardigrade claw structure
(1) Internal claw, (2) external claw and (3) base. Redrawn from Bertolani (1982b).

shape and symmetrical with respect to the median plane of the leg. While in the Hypsibiidae, the two double claws on each leg are usually of different size and shape and asymmetrical with respect to the median plane of the leg. In the Apochela, the primary and secondary branches are both attached directly to the leg (see Figure 9.8).

In some species, a cuticular thickening (lunule) surrounds the base of each double claw. These vary in size and shape and may have a smooth edge or may be dentate (Figure 3.13). Additional cuticular bars may also be observed in some species, below the base of the claws or beside the internal claw (Figure 3.12). The presence or absence of such bars has been considered enough to separate species in some instances (e.g. Ito and Tagami, 1993).

The cuticle may be smooth, granulated or bearing tubercles and may or may not have pores. The cuticular extensions of the Eutardigrada are less exaggerated than those of the Heterotardigrada and are limited to spines or humps (sometimes termed gibbosities). Humps are particularly common in members

External morphology

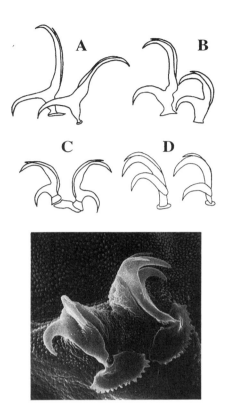

Figure 3.13. Examples of eutardigrade claws
(A) *Isohypsibius*-type, (B) *Hypsibius*-type, (C) *Dactylobiotus*-type, (D) *Eohypsibius*-type. SEM shows *Macrobiotus richtersi* claws showing the well-developed dentate lunules. SEM courtesy of Hieronim Dastych.

Figure 3.14. Lateral view of *Calohypsibius ornatus*
SEM courtesy of Diane Nelson. Bar ≈ 50µm.

of the genus *Isohypsibius*, while spines occur in rows across the dorsum of some species such as *Calohypsibius ornatus* (Figure 3.14).

While the pharyngeal apparatus is usually visible through the transparent cuticle of most eutardigrades, it is strictly part of the internal anatomy and is described in detail in Chapter 4. However, the buccal aperture exhibits a number of external structures which may be of help in taxonomy. These structures include lobes (typically 6 in many genera), papullae (10 in *Minibiotus* and *Haplomacrobiotus*; 6 in *Calohypsibius*) and lamellae (30 in *Pseudobiotus*; 14 in *Amphibolus*; 12 in *Thulinia* and 10 in most Macrobiotidae) (Figure 3.15). In the mouth of many eutardigrade genera, an additional buccal armature of bands of small teeth (mucrones) or indentations can be observed (Figure 3.15)

The cuticle of a number of eutardigrade species displays coloration resulting from pigmentation of the epidermal cells. Although this is a variable

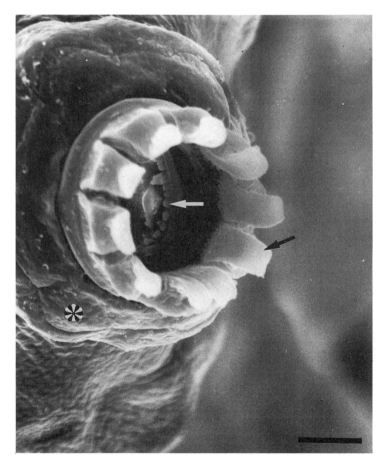

Figure 3.15. Peri-buccal region of *Macrobiotus tonollii*, showing lamellae (black arrow), mucrones (white arrow) and circum-oral sensory field (*)
SEM courtesy of Diane Nelson. Bar = 5 µm.

characteristic, not considered to be of great value in the descrimination of species, it is such a prominent characteristic of some fresh specimens that it acts as an instant guide suggesting possible affinities to groups of species. For example, a number of species in the *oberhaeuseri* group exhibit distinctive red/brown stripes on the dorsum (see Chapter 9). Coloration cannot be used to compare preserved specimens as it fades rapidly in mounting media such as Hoyer's and is, therefore, of little taxonomic value.

4

Internal structure

Internal arrangement

Despite the wide range of aquatic environments colonized by tardigrades, from hot springs to oft-frozen polar pools and from immense oceans (offering great environmental stability) to small temporary pools of water, the group does not display a great variation of internal organ structure and organization. Exceptions to this are the feeding apparatus and cuticle, which are markedly different in the two Classes, Heterotardigrada and Eutardigrada. In addition, the excretory organs of eutardigrades (Malpighian tubules) are absent in heterotardigrades. The midgut also differs between the two classes. The heterotardigrade midgut has five or six lateral diverticula, while that of eutardigrades is straight.

Intestine

The intestine is often visible in living eutardigrades, but obscured by the cuticular plates in echiniscids. The alimentary canal consists of five structurally distinct regions: buccal tube, pharynx, oesophagus, midgut and hindgut (Figure 4.1). The first three sections collectively constitute the foregut. The foregut and hindgut are shed during moulting. The sclerified structures of the pharynx are usually most conspicuous, but the midgut is also often highlighted by its coloured contents in species feeding on chlorophyllous material or other pigmented food items.

Near the anterior end of the buccal tube, dorsal and ventral anterior apophysies act as anchors for the stylet muscles. The shapes of these apophyses vary greatly between genera, and are described as crest-shaped or as semilunar hooks (Figure 4.2). The buccal tube ends in three thickenings at the proximal end, the bulb apophyses — two lateral apophyses and one larger dorsal apophysis (Figures 4.2 and 4.3a), which alternate in position with the bulb placoids (Figure 4.4). These are excluded from measurements of the buccal tube.

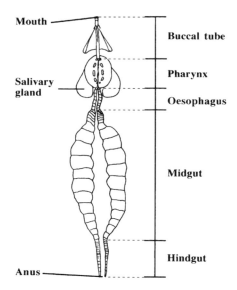

Figure 4.1. Dorsal view of the eutardigrade alimentary canal showing the five main morphological regions
Redrawn and simplified from Raineri (1985).

Apophyses are absent in *Milnesium*, but cuticular flaps at the proximal end of the buccal tube are probably homologous structures (Dewel and Clark, 1973a). The walls of the buccal tube separate in the region of the apophyses as the buccal tube enters the pharynx (Figure 4.3b), resulting in a flaring of the proximal end of the buccal tube in some species.

The pharynx and its associated apparatus forms a complicated structure used for feeding (Figure 4.2). A muscular pharyngeal bulb (typically about 20–30 µm in diameter) is found surrounding the proximal end of the buccal tube. The bulb is a simple ovoid or sometimes pear-shaped, but it is never an elongated multi-staged pump as is characteristic of many nematodes. The bulb is separated from the body cavity by a basal membrane (Baccetti and Rosati, 1969) and is separated from its lumen by a cuticular layer which is strengthened by bars (placoids) in most species. These placoids occur in rows, each consisting of three pairs. For clarity, drawings often only show two pairs, and the elements of each pair are so close together that a false impression is given that each row of placoids consists of two elements rather than six (Figures 4.2 and 4.4). Some placoids have a constriction (throttling) along their length. The lumen of the bulb is tri-radiate in transverse section, as is the proximal end of the buccal tube (Figure 4.2.5). In some species within the genus *Diphascon*, a septulum is present posterior to, and alternating with, the placoids (Figure 4.4). Placoids are reduced or absent in *Itaquascon* spp. and absent in *Milnesium* spp. and *Limmenius* spp.

A pair of stylets occurs anterior to the pharyngeal bulb, on either side of the buccal tube. These are rigid structures, lying in the lumen of the salivary glands, which are pushed out through the buccal cavity by a series of muscles in order to pierce the cells of food items and allow their fluid contents to be ingested. The stylets are guided into the buccal cavity by two lateral flanges

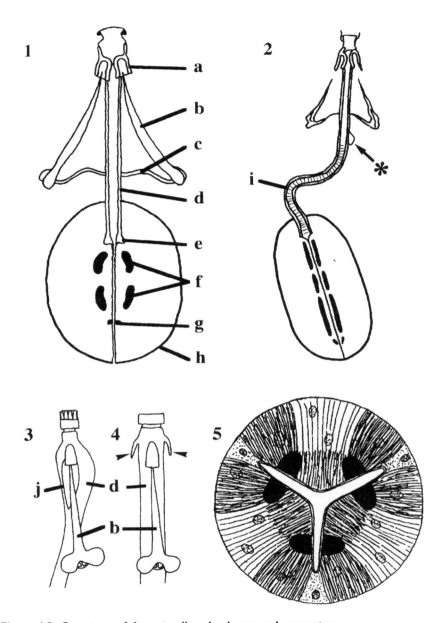

Figure 4.2. Structure of the eutardigrade pharyngeal apparatus
(1) Dorsal view of the *Hypsibius/Macrobiotus*-type apparatus: a, lateral flange; b, stylet; c, stylet support; d, buccal tube; e, bulb apophysis; f, macroplacoid; g, microplacoid; h, pharyngeal bulb. (2) Dorsal view of the *Diphascon/Pseudodiphascon*-type apparatus: i, flexible portion of buccal tube; *, teardrop-shaped thickening used to distinguish the *Diphascon* subgenera, *Diphascon* and *Adropion* (see text). (3) Lateral view of the *Macrobiotus/Pseudodiphascon*-type buccal tube: j, ventral supporting bar. (4) Lateral view of the *Hypsibius/Diphascon*-type buccal tube: arrow heads indicate the dorsal and ventral apophyses for the insertion of the stylet muscles. (5) Transverse section of a typical pharyngeal bulb showing the tri-radiate lumen. Redrawn from Dastych (1992), Greven (1980), Pennak (1989) and Ramazzotti and Maucci (1983).

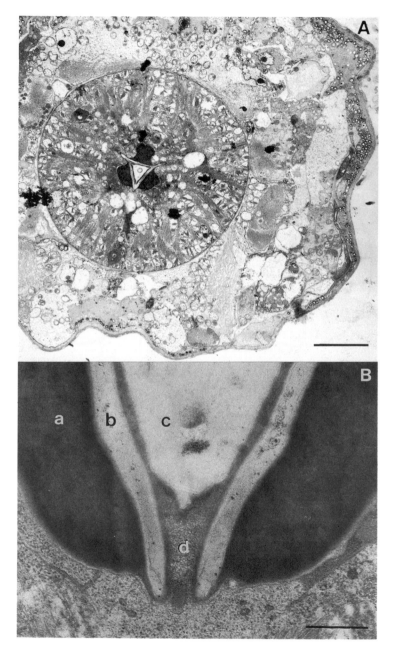

Figure 4.3. TEMs of pharyngeal bulb in transverse section through the region of the bulb apophyses in *Ramazzottius varieornatus*
(A) View of entire section (bar = 10 μm). (B) Detail of lateral bulb apophyses: a, lateral apophysis; b, wall of buccal tube; c, lumen of buccal tube; d, gap between the separated walls of the buccal tube. (Bar = 1 μm.)

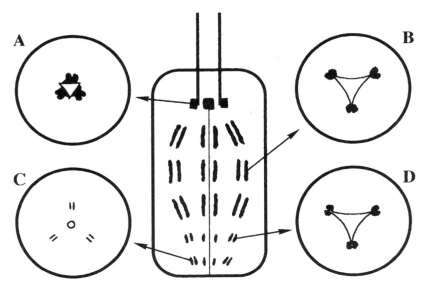

Figure 4.4. Diagrammatic representation of a pharyngeal bulb, flattened during preparation to show the separation of the placoids
A–D, Transverse sections of the bulb to show the symmetry of the sclerified structures: A, apophyses; B, macroplacoids; C, septula; and D, microplacoids. It can be seen that A and C alternate with B and D.

(termed 'stylet sheaths' in *Milnesium tardigradum*). Suction is provided by contraction and relaxation of the arrays of radial muscles of the pharyngeal bulb which extend from the periphery of the bulb to the surface of the lumen (about 15 μm in length). The length of the stylets may be straight or curved; the proximal end (the furca) usually displays a bifurcation, onto which are inserted the stylet muscles. The position of the stylets is maintained by two flexible stylet supports which articulate with the stylet furca at one end and are inserted upon the buccal tube at the other. The pharyngeal apparatus, therefore, offers a number of structures which are of great taxonomic value, including the size, orientation and symmetry of the anterior stylet apophyses and the size and number of rows of placoids.

Four basic types of pharyngeal apparatus are recognized in parachelate eutardigrades. The *Hypsibius* and *Macrobiotus* types have a rigid buccal tube (Figure 4.2.1); the *Diphascon*, and *Pseudodiphascon* types have a flexible, annulated posterior portion of the buccal tube (Figure 4.2.2); *Macrobiotus* and *Pseudodiphascon* have a ventral supporting rod on the buccal tube (Figure 4.2.3) which is absent in *Hypsibius* and *Diphascon* (Figure 4.2.4). A number of genera have a pharyngeal apparatus based on the *Diphascon* model (e.g. *Diphascon, Hebesuncus, Mesocrista* and *Platicrista*) which are distinguished by the shape of the apophyses for the insertion of muscles on the stylets (Pilato, 1987). Within the genus *Diphascon*, Pilato has identified two subgenera; *Diphascon (Diphascon)* which has a teardrop-shaped cuticular lump (Figure 4.2.2) at the junction of the buccal tube and the pharyngeal tube (e.g. *Diphascon* (*Diphascon*) *chilenense*-type species) and *Diphascon* (*Adropion*) which lacks the teardrop-shaped structure (e.g. *Diphascon* (*Adropion*) *scoticum*). This teardrop-shaped structure has been termed the 'posteriodorsal apodeme' by Dastych (1992).

The light microscope observations of Marcus (1929*b*) showed that the bulb of those species studied consisted of 51 cells arranged in hexads of radial and inter-radial muscle and epithelial cells. The 27 epithelial cells produce the placoids and the basal membrane, while the 24 muscular cells generate the suction power of the organ. The structure of the pharyngeal bulb of the apochelate *Milnesium tardigradum* exhibits a number of differences to that of the Parachela (Marcus, 1929*b*; Dewel and Clark, 1973*a*) (Figure 4.5). The bulb is pear-shaped rather than spherical or ovoid and placoids are absent. In general, the pharyngeal apparatus of heterotardigrades is of less taxonomic importance than that of the eutardigrades. It typically exhibits thin straight stylets and a pharyngeal bulb with cuticular bars (bulbar rods) rather than discrete placoids (Figure 4.6).

The eutardigrade oesophagus is lined with small cuboidal cells which would seem to have a secretory function (Dewel and Clark, 1973*b*). The mucoid secretions of these cells probably function in lubrication or protection of the apical surfaces of the midgut cells. Alternatively, the secretion could activate digestive enzymes produced by other organs such as the salivary glands.

The structure of the eutardigrade midgut has been studied in detail by Greven (1976). The sac-like organ is composed of epithelial cells organized into a convoluted monolayer. Only a single cell type is present, but it displays considerable morphological variation. Cytosis vesicles, mitochondria and

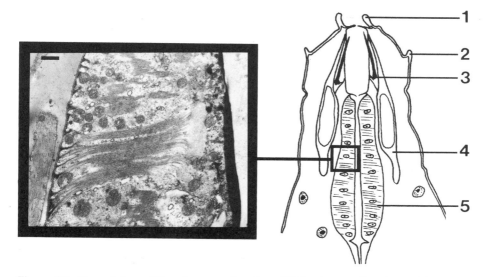

Figure 4.5. Dorsal view of the pharyngeal region of *Milnesium tardigradum*, with (inset) TEM of the muscle fibres of the pharyngeal bulb
(1) Oral papilla; (2) lateral papilla; (3) stylet; (4) salivary gland; and (5) pharyngeal bulb. Bar = 2 μm.

Figure 4.6. Pharyngeal apparatus of Batillipes sp.
Redrawn from McKirdy (1975).

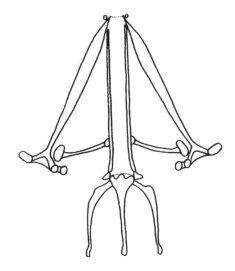

abundant rough endoplasmic reticulum are characteristic of the cells — many are also filled with large amounts of storage materials, particularly polysaccharides and lipids. Observations of granules released from the midgut cells into the lumen and expelled with the faeces, has led to the suggestion that the midgut has a secondary excretory function (Marcus, 1929b; Greven, 1976). Cell surfaces in the lumen of the midgut have been observed to be smooth in some regions and covered with slender microvilli (approximately 0.25 µm in length) in others (Greven, 1976). The occurrence of a well-developed rough endoplasmic reticulum has been taken as morphological evidence of enzyme synthesis (Greven, 1976) and has been supported by subsequent histochemical observations (Raineri, 1985). The midgut of some species has also been observed to support an extensive bacterial flora (see Chapter 7).

Cuticle

Studies of the cuticle of tardigrades have attracted considerable interest because of its alleged phylogenetic significance and the long-disputed association of the Tardigrada with the Arthropoda (e.g. Baccetti and Rosati, 1971; Bussers and Jeuniaux, 1973) and also because of its presumed involvement in the process of anhydrobiosis (e.g. Greven and Greven, 1987; Wright, 1988a; 1989a).

The structure of the cuticle varies considerably between the two orders (e.g. Greven, 1972). Within the Eutardigrada, although the ultrastructure shows considerable variation between species (e.g. Crowe et al., 1971; Baccetti and Rosati, 1971; Bussers and Jeuniaux, 1973; Greven, 1975; 1983; Wright, 1988a; 1988b; Węglarska, 1989a), there are a number of generalizations that can be made for the group. The cuticle has several discernible layers. An outer epicuticle, a lipid-rich intracuticle and a procuticle (Figure 4.7). The surface of the epicuticle may be covered in a flocculent mucous coat and the epicuticle and

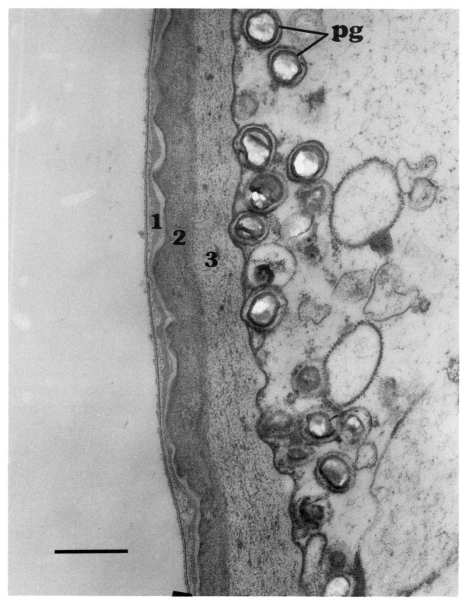

Figure 4.7. TEM of section of eutardigrade intrasegmental cuticle
(1) Epicuticle; (2) intracuticle; and (3) procuticle. Abbreviations used: pg, pigment granules. (Bar = 0.5 μm.)

intracuticle are separated by a trilaminate layer. An epidermal layer lies beneath the procuticle. The epidermis may contain pigment granules particularly in the intrasegmental areas where the cuticle tends to be thickest.

The absolute thickness of the various cuticular layers is difficult to assess accurately as sectioning angles are difficult to determine. The animals invariably experience some distortion during preparation for electron microscopy and, as can be seen from Figure 4.7, significant variation in thickness exists even in adjacent areas so that average figures are of limited value. Wright (1988a) explains the problems of taking measurements from thin sections and provides a method for calculating cuticular dimensions from numerous random sections — assuming the tardigrade body to be equivalent to a squat cylinder. Though highly variable, the total cuticle is typically about 0.5 μm in thickness with the component layers varying as shown in Table 4.1.

The structure of the heterotardigrade cuticle is quite different to that of the eutardigrades. The epicuticle is typically composed of supporting pillars (Figure 4.8). These are quite distinct structures in the ventral cuticle, becoming

Table 4.1. The thickness of cuticular sublayers in eutardigrades

	Thickness		
	Minimum	Maximum	Typical
Epicuticle	0.04 μm	0.6 μm	0.1 μm
Intracuticle	0.1 μm	0.26 μm	0.2 μm
Procuticle	0.2 μm	2.0 μm	0.25 μm

After various authors and unpublished observations.

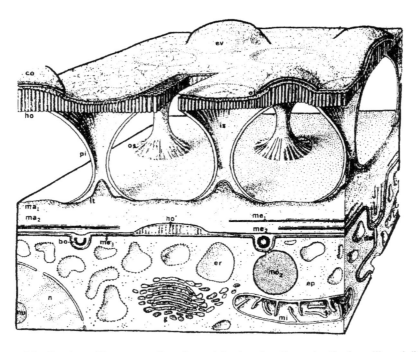

Figure 4.8. Section of heterotardigrade cuticle showing the epicuticular pillars (pi) supporting the striated honeycomb layer (ho)
Reprinted from Kristensen (1976), with permission.

more densely packed in the dorsal cuticle where the pillars seem to coalesce to form a reticulate structure with numerous air-filled lacunae (Figure 4.9). Pillars are absent in regions near claws, sensillae and muscle attachments (Kristensen, 1978). The pillars support a striated (honeycomb) layer which is composed of numerous hexagonal subunits. Epicuticular pillars are considered to be a heterotardigrade characteristic, but they have also been observed in a few freshwater eutardigrade species in the genus *Dactylobiotus* (Greven, 1982; Weglarska, 1989*a*) and an aberrant *Macrobiotus* sp. from a hot spring (Kristensen, 1982*b*). In addition, the cuticle of the heterotardigrade *Echiniscoides sigismundi* is more like that of many eutardigrades than the other heterotardigrades so far studied (Greven and Grohé, 1975). The usual descriptions of eutardigrade- and heterotardigrade-type cuticular structure may, therefore, have limited taxonomic value.

Muscle

The somatic muscles of tardigrades are attached to the cuticle by a series of discrete structures. One or more muscles may be attached at each point (Figure 4.10). Locomotor activity is controlled by the somatic muscles which flex the body and legs against the antagonistic hydrostatic pressure of the body cavity fluid.

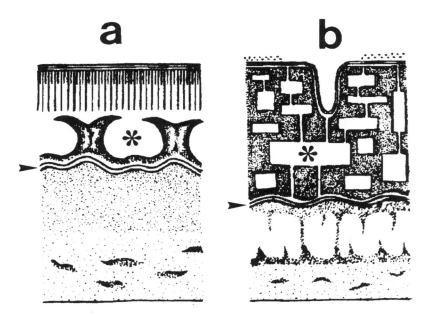

Figure 4.9. Variation between the ventral (a) and dorsal (b) cuticle structure in *Echiniscus testudo*
The pillar layer, immediately above the trilaminate layer (arrow head), exhibits discrete pillars in the ventral cuticle which coalesce in the dorsal cuticle. Air-filled lacunar spaces (*) are a feature of this layer in both regions. Modified from Greven (1984).

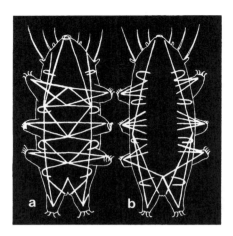

Figure 4.10. The arrangement of the somatic muscles and their cuticular attachments in *Echiniscus* sp. Redrawn from Marcus (1929b).

A thickening of the procuticular layer to provide strength and support can often be observed at these muscle insertions (Figure 4.11), with conical projections extending into the epidermis (Shaw, 1974).

There have been few ultrastructural observations of the somatic muscles of eutardigrades. Shaw (1974) and Walz (1973; 1974; 1975a) have both examined the muscles of *Milnesium tardigradum* and *Macrobiotus hufelandi* and it is generally assumed that little structural variation exists within the Eutardigrada. The somatic muscles have been considered as smooth muscle by Marcus (1929b) and Hanson and Lowy (1960). More recently, the contractile apparatus has been shown to display characteristics of both smooth and striated muscle and is considered an intermediate type, structurally similar to that of *Peripatus* (Walz, 1974; 1975a). Kristensen (1978) has shown that both smooth and striated muscle occurs in tardigrades, but the latter is predominantly found in the most primitive species — marine heterotardigrades. Kristensen has suggested that a partial shift from arthropod-like striated muscle to a smooth type of muscle might have accompanied a transition from the marine to the terrestrial environment, and might somehow be functionally tied to the development of anhydrobiosis in terrestrial species (Kristensen, (1978).

In transverse section, two main arrangements of muscle filaments have been observed in eutardigrade somatic muscles (Shaw, 1974; Walz, 1974).

- Thick filaments occurring in clumps, separated by clumps of thin filaments. A few dense bodies are also present in this arrangement.
- More commonly, thick filaments are arranged in regular arrays, each surrounded by between five and seven other thick filaments at an average distance of about 50 nm. The thin filaments are interposed between six and 12 thin filaments around each thick filament. Among the filaments are a number of dense bodies (possibly composed of aggregations of thin filaments), each surrounded by an electron-translucent 'halo' in which there are no filamentous elements. This arrangement is shown in Figure 4.12.

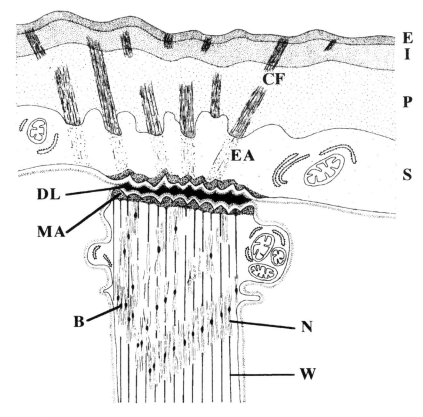

Figure 4.11. Structure of a cuticular muscle attachment
E, Epicuticle; I, intracuticle; P, procuticle; S, epidermis; CF, cuticular fibres; EA, intra-epidermal attachment; DL, dense layer; MA, intramuscular attachment material; B, dense body; N, thin filament; W, thick filament. Modified from Greven (1980).

The diameters of the somatic muscle filaments in *Macrobiotus hufelandi* have been calculated by Shaw (1974) and are shown in Table 4.2.

Nervous system and sensory organs

The tardigrade nervous system has been studied by Marcus (1929b), Greven and Kuhlmann (1972), Walz (1975b; 1978; 1979b) and Kristensen (1981). It is built on the annelid/arthropod plan — displaying distinct metamerism. The system consists of a dorsal lobed cerebral ganglion (also known as the brain) and a ventral sub-oesophageal ganglion. A chain of four bi-lobed ventral ganglia (serving the four pairs of legs) is joined by a pair of ventral nerve cords (Figure 4.13).

The nervous system of the marine eutardigrades in the genus *Halobiotus*, is very different to that described for other genera. The brain is very large, consisting of three lobes which may be homologous with the protocerebrum,

Internal structure   51

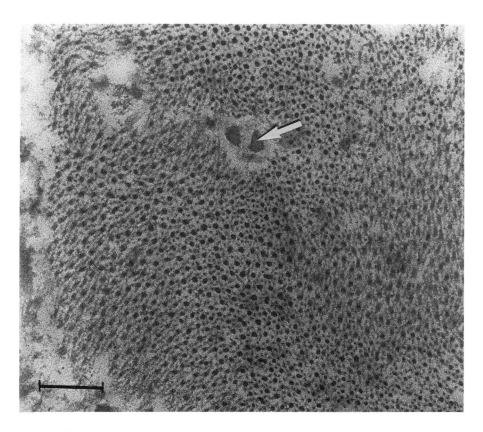

Figure 4.12. TEM transverse section of eutardigrade somatic muscle showing the typical arrangement of thin and thick filaments, and dense bodies (arrow) Bar = 0.1 µm.

Table 4.2. Diameters of muscle filaments in *Macrobiotus hufelandi*

	Diameter	
	Average	Range
Thick filaments	16 nm	13–22 nm
Thin filaments	5 nm	3–8 nm

deutocerebrum and tritocerebrum of the eu-arthropods (Kristensen, 1982a). The phylogenetic significance of these observations is, as yet, unclear.

The structure of the sensory receptors in the cirri and clava of the marine arthrotardigrades has been described by Kristensen (1981) as homologous to that of arthropod sensillae (Figure 4.13). The cuticle extending over the cirri and clava of echiniscids has been shown to be very dissimilar (Dewel *et al.*, 1993), reflecting differences in function. In *Echiniscus viridissimus*, the cuticle over the cirri is similar to that typically found on the ventral surface of the animal. The

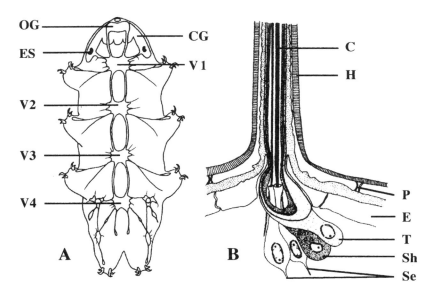

Figure 4.13. General plan of the nervous system of a eutardigrade (A) and details of a heterotardigrade cephalic sense organ (B)
Abbreviations used: OG, oesophageal ganglion; CG, cerebral ganglion; ES, eyespot; V1–V4, ventral ganglia 1–4; C, cilium; H, honeycomb layer of cuticle; P, cuticular support pillar; E, epidermis; T, Trichogen cell; Sh, sheath cell; Se, sensory cell. (A) Modified from Marcus (1992b), after several authors. (B) Reprinted from Kristensen (1981), with permission.

cuticle over the clava, however, lacks an intra- and an endocuticle, and the epicuticular pillar layer is filled with anastomosing branches from a perforated base — similar to that typical of the dorsal echiniscid cuticle (Figure 4.9).

Eutardigrades generally lack the sensory appendages that are such a feature of the heterotardigrades. Internal cephalic structures reported by Walz (1978) in *Macrobiotus hufelandi*, are presumed to have sensory functions and are described as 'sensory fields'. The cuticle extending over sensory areas has been shown to exhibit an altered permeability to allow the diffusion of stimulating substances through the cuticle to underlying cells that are presumed to be chemoreceptive (Walz 1975b; 1978). A pair of cup-shaped pigmented eyespots associated with the lateral lobes of the cerebral ganglion is also present in many species.

Excretory organs

Excretion is likely to occur in up to four different ways: (i) Through the action of salivary glands during ecdysis when the buccal apparatus is expelled through the buccal cavity. (ii) Via the cuticle when it is shed during ecdysis. The cuticle accumulates excretory granules between moults. (iii) Through the wall of the midgut. (iv) Through specialized excretory glands (Malpighian tubules) into the hindgut in eutardigrades and, possibly, through ventral organs into the cuticle in heterotardigrades.

Eutardigrades have three Malpighian tubules: one lying dorsal to the midgut and two lying laterally. Each organ is composed of a distal and a proximal segment with each segment composed of three cells. These cells can be variously arranged, linearly as in *Macrobiotus richtersi* (Węglarska 1980; 1989b) or as a trefoil in *Hypsibius microps* (Figure 4.14) and *Isohypsibius granulifer* (Węglarska 1987a; 1989b). The organs are identical in both sexes. An internal canal passes through the middle of the organ. The distal segment probably produces a primary urine by active transport and ultrafiltration. The proximal segment then modifies the urine before it passes into the lumen of the pylorus — the junction between the midgut and the hindgut (Dewel and Dewel, 1979; Greven, 1979; Węglarska, 1980; Dewel et al., 1993).

Milnesium tardigradum has four excretory organs (one dorsal, two lateral and an additional ventral organ) in which the distal and proximal segments are each composed of six cells (Dewel and Dewel, 1979). Small muscles inserted on the pylorus are thought to allow this structure to operate as a valve controlling loss of material from the midgut and excretory organs (Dewel and Dewel, 1979).

The size of the excretory organs in eutardigrades depends on the environment in which the animals live. Bryophilous species have larger organs than their freshwater relatives (Węglarska, 1989b) and, uniquely for a marine tardigrade species, the eutardigrade *Halobiotus* exhibits particularly large Malpighian tubules, suggesting that it is only secondarily adapted to the marine environment (Crisp and Kristensen, 1983). Marine heterotardigrades are isotonic with sea water and so require no osmoregulatory organs, whereas eutardigrades are anisotonic with fresh water and so require specialized mechanisms to control water uptake through the cuticle.

Malpighian tubules are absent in heterotardigrades, though structures have been observed which are thought to perform osmoregulatory and excretory functions. These ventral organs (composed of one median and two lateral cells enclosing a pair of convoluted canals) have been observed at the level of the

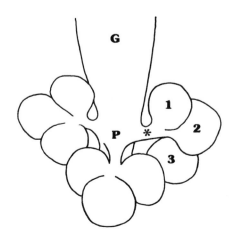

Figure 4.14. Dorsal view of the excretory organs of *Hypsibius microps*
G, midgut; P, pylorus; *, proximal segment of the right lateral Malpighian tubule; 1–3, the three cells of the distal segment the right lateral Malpighian tubule. Redrawn from Węglarska (1989b).

second and third pairs of legs in species of *Echiniscus, Hypechiniscus* and *Pseudechiniscus* (Dewel *et al.*, 1992). These organs are probably transporting material accumulated in the body cavity fluid to the endocuticle, and from here via cuticular pores and lacunae to the outer cuticular layers or to the surface. It is not known whether this process is primarily excretory or secretory. In addition, the organs seem to be involved in the translocation of bryophyte metabolites and may, therefore, represent a major evolutionary innovation that facilitated the invasion of terrestrial microhabitats by the Echiniscidae (Dewel *et al.*, 1992).

Body cavity cells

The body cavity cells (also known as coelomocytes) are often clearly visible as they move passively in the body cavity of live eutardigrades, transported by currents in the body cavity fluid caused by the animal's movements (Figure 4.15). The cells (in the region of 200 per animal) are often seen particularly clearly in the caudal region where they are not obscured by other internal organs. Using both light and electron microscopy, the overall shape of body cavity cells has been shown to be quite variable; those in contact with the epidermis or intestine have irregular edges, while others have a more regular ovoid shape with smooth edges (May, 1946; Rosati, 1968; Węglarska, 1975; Kinchin, 1993). Body cavity cells are metabolically very active and typically have a large conspicuous nucleus. In addition, the cytoplasm is packed with other organelles and storage vacuoles containing reserves of lipids and carbohydrates which are thought to be important in promoting anhydrobiotic survival (see Chapter 6). The surface irregularities observed in some cells are thought to indicate that materials pass between the body cavity cells and those of the alimentary canal or epidermis, though an excretory function for these cells has been ruled out by Rosati (1968). The main function of the cells is to store food reserves and, therefore, their size and number are thought to be indicative of the nutritional condition of the animal. In addition, the recently observed tyrosinase activity in the body cavity cells of *Macrobiotus hufelandi* suggests a defence role for these cells — similar to that mediated by insect haemocytes (Volkman and Greven, 1993). A phagocytic function had previously been proposed for the body cavity cells by Greven (1980) who observed bacteria in the body cavity cells of infected animals.

In the heterotardigrades, the body cavity is proportionally smaller than that of the eutardigrades. In addition, the body cavity cells of the echiniscids appear to be anchored to the gut and/or body wall (Dewel *et al.*, 1993).

Internal structure

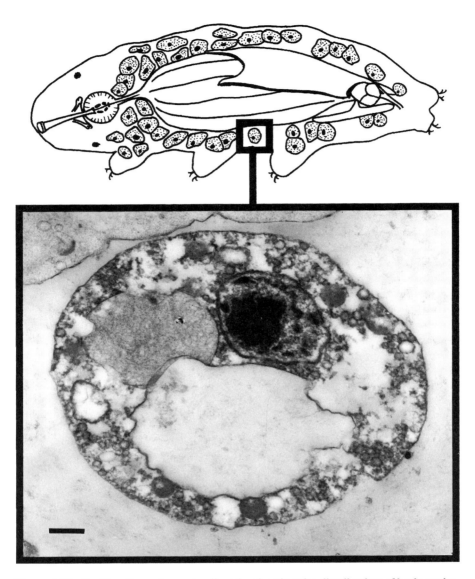

Figure 4.15. Section through a eutardigrade, showing the distribution of body cavity cells with (inset) TEM of a section of a body cavity cell
Reproduced from Kinchin (1993), with permission. (Bar = 1 μm.)

5

Reproduction and life history

Mating

In gonochoristic species, mating can take place between a female and a single male or several males with the males clinging to the anterior part of the female with their front legs. According to Bertolani (1992) this grip is so tenacious that the animals will not separate when disturbed or even placed in mounting medium. In limno-terrestrial eutardigrade species, fertilization usually occurs inside the body of the female (Bertolani, 1990), whereas marine species employ external seminal receptacles (see Chapter 3).

Although the mechanism of mate location is not understood, the cephalic sensory structures are thought to have a role. Female heterotardigrades are actively sought by males, the male possibly stimulating the female by tactile use of the lateral cirri (Kristensen, 1981). Bertolani (1992) has observed that in the eutardigrade *Pseudobiotus megalonyx*, diploid males were not only capable of locating a mate, but were also able to distinguish between two morphologically similar types of females — 'desirable' diploid amphimictic individuals mixed with 'undesirable' triploid parthenogenetic individuals.

Hermaphroditism

In hermaphrodite species, male and female gametes are produced by the same individual — in a single, syngonic, gonad in eutardigrades such as *Isohypsibius granulifer* (Węglarska, 1987). Therefore, hermaphrodites have a two-fold advantage in reproductive capacity over gonochoristic species, since every individual in the population is able to lay eggs. Hermaphroditism is usually considered to be a primitive condition in animals and to have given rise to amphimictic species. However, hermaphrodites are unknown in primitive heterotardigrades. This form of sexuality is only known in the more advanced parachelate eutardigrades, including species of *Macrobiotus*, *Amphibolus*, *Isohypsibius*, *Parhexapodibius* and a single marine heterotardigrade genus,

Orzeliscus (Bertolani, 1979*b*; 1987*a*; Bertolani *et al.*, 1983; Bertolani and Manicardi, 1986), suggesting that it is a derived character that has developed late in the evolutionary history of tardigrades. Hermaphroditic tardigrades, therefore, arose from amphimictic species. A similar picture has been described for the nematodes (Triantaphyllou and Hirschmann, 1964), suggesting a parallel evolutionary sequence in the two groups.

Specimens of several hermaphrodite tardigrade species have been observed with both spermatozoa and near-mature oocytes in a single ovotestis, making the option of self-fertilization (automixis) a possibility. There are some advantages to self-fertilization. As in parthenogenesis, self-fertilizing hermaphrodites have complete independence from mating contacts, but more genetic variation is achieved than in ameiotic parthenogenesis by allowing crossovers during meiosis and the independent assortment of chromosomes. However, self-fertilization results in lower fitness than amphimixis because of increased homozygosity, allowing the expression of disadvantageous recessive genes.

All the populations of bryophilous heterotardigrades examined by Bertolani and Manicardi (1986) were gonochoristic or unisexual — hermaphrodites were not observed. Whereas species of *Pseudechiniscus*, *Bryodelphax* and *Hypechiniscus* are often bisexual, species of the genus *Echiniscus* were, until recently, thought to be exclusively thelytokous. Observed variations in the gonopore were attributed to alterations following oviposition (Nelson, 1982*a*). However, re-examination of material by Dastych (1987*b*) has confirmed that males exist in a number of *Echiniscus* spp.

Reproductive organs

Differences in reproductive anatomy are apparent between the two classes of tardigrade. In the Eutardigrada the paired sperm ducts of the male and the single oviduct of the female open into the rectum. The eutardigrade anus can, therefore, be considered a true cloaca. In the Heterotardigrada, the sperm ducts and oviduct open into a separate ventral, pre-anally situated gonopore, often surrounded by a rosette of cuticular folds (Figure 5.1). Exceptions to this are *Lepoarctus coniferus*, *Tetrakentron synaptae* and *Tholoarctus natans*, all of which exhibit a post-anal gonopore (Kristensen, 1980; Kristensen and Renaud-Mornant, 1983).

Gonoporal dimorphism has been described in the marine arthrotardigrades by Pollock (1970*a*; 1975*a*). An oval, slightly raised, posteriorly positioned gonopore is associated with males, while a rosette of plates typically surrounds a more anteriorly positioned female gonopore. The female gonopore is flanked by two seminal vesicles or 'annex glands' (Pollock 1970*a*), each consisting of a ventral vesicle and a genital duct whose exterior opening is anterior to the gonopore (e.g. Renaud-Mornant, 1970; 1979; 1982*a*; Grimaldi de Zio *et al.*, 1982*c*; 1990*a*; Noda, 1987; Kristensen and Higgins, 1984*a*; 1984*b*; 1989). Spermatozoa have been observed in the ducts of *Renaudarctus psammocryptus*

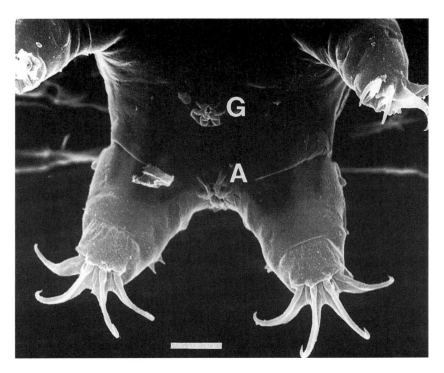

Figure 5.1. SEM of *Echiniscus* sp. caudal region (ventral view)
(G) Gonopore; (A) anus. SEM courtesy of Diane Nelson. (Bar ≈ 10 μm.)

(Kristensen and Higgins, 1984*a*). A similar gonoporal dimorphism has been observed in terrestrial heterotardigrades. In bisexual populations of *Pseudechiniscus* spp., *Bryodelphax* spp. and *Hypechiniscus* spp., dimorphism of the gonopore has been observed (six-petalled rosette in the female; small protruding tube in the male). A seminal receptacle is also present during the autumn and winter in some female eutardigrades belonging to the genera *Macrobiotus* and *Hypsibius* (Marcus, 1928). This sac-like structure opens into the hindgut, near the oviduct, and immobile spermatozoa have been observed in the receptacle (Nelson, 1982*a*). In *Milnesium tardigradum*, the distal segment of the ventral Malpighian tubule seems to have been modified as a reproductive gland. This is considered by Dewel *et al.* (1993) to be an example of exaptation (Gould and Vrba, 1982).

Sex reversal in heterotardigrades has also been hypothesized following an observation of specimens of *Batillipes pennaki* apparently displaying female gonads, but male gonopores (Grimaldi de Zio and D'Addabbo Gallo, 1975). However, this observation has not been repeated.

Sexes can be distinguished within the genus *Halechiniscus* by the relative lengths of their cephalic appendages. In males, clava are longer than the adjacent lateral cirri; in females, they are shorter (Schulz, 1955). Secondary sexual characteristics in eutardigrades are more or less confined to minor differences of the claws (Bertolani, 1992). Typically, males display an extra spur or hook on the inner claw of the first pair of legs (e.g. *Milnesium tardigradum*, *Pseudobiotus megalonyx*). Species within the genus *Ramazzottius* display a lateral bulge on the fourth pair of legs — flattened in the male and rounded in the female (Baumann, 1966; Rebecchi and Bertolani, 1988). The function of this structure is unclear. It is suspected that secondary sexual characteristics (which are not observed in juveniles) arise at the moult immediately preceding mating and egg-laying (Bertolani, 1992).

A single sac-like gonad occurs in a position dorsal to the midgut (Figure 5.2). The gonad in both sexes is suspended from its anterior end by ligaments, inserted on the cuticle above the pharynx. The gonad exhibits a capacity to change size according to its developmental state. In many females, the ovary alternates with the midgut to occupy the internal body cavity, thus allowing the animal to maintain a small body size (Bertolani, 1983*b*). The animal, therefore, oscillates between periods of feeding and periods of reproduction. A hormonal control mechanism for gonad maturation, exerted by neuroendocrine cells of the cerebral ganglion, has been hypothesized by Raineri (1987).

Reproductive modes

Species are sometimes differentiated according to the life history strategy they adopt. Some optimize fitness by maximizing their intrinsic rate of increase (r-selected species), while others maximize their share of the carrying capacity of the habitat (K-selected species) as described by Pianka (1970) and Parry (1981). While r and K represent extremes on a continuum, most species will show a mixture of r and K influences in the environment. Extreme r-selection favours reproduction at the expense of adult survival — semelparity, reproduction at a small body size and high fecundity. This is typical of pioneer species and those that regularly experience high density-independent mortality owing to environmental fluctuations. Extreme K-selection, on the other hand, favours

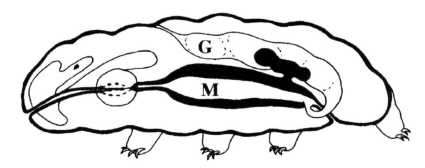

Figure 5.2. Longitudinal section of a eutardigrade showing the position of the gonad (G) dorsal to the midgut (M)

survival of the parent rather than a high reproductive capacity — iteroparity, large adult size and relatively low fecundity. These organisms experience density-dependent mortality.

In general, tardigrades tend towards r-selection. However, tardigrades are known to exhibit various types of sexuality and reproduction with populations consisting of males and females, females alone or hermaphrodites. One might expect the species inhabiting more-constant environments to be more K-selected than those inhabiting more-fluctuating environments. In fact, marine tardigrade species tend to be gonochoristic (and possibly iteroparous), while parthenogenesis (of particular benefit to pioneer species) is confined to limno-terrestrial species. In a study of reproductive cycles in three species of gonochoristic freshwater and terrestrial eutardigrades, Rebecchi and Bertolani (1992) showed that while all the females studied were iteroparous, males could be iteroparous or semelparous. These differences reflect systematic position and habitat differences. Evidently, more species will need to be examined for evidence of r- or K-selection before detailed patterns can be identified regarding tardigrade reproductive strategies. While there have been some criticisms of the use of the r–K model (e.g. Roff, 1992; Stearns, 1992), it provides a useful initial generalization when contrasting life histories.

Parthenogenesis

Although several cytological mechanisms for parthenogenesis have been described in thelytokous populations, ameiotic parthenogenesis (apomixis) is the most common in tardigrades (Bertolani, 1982a; Rebecchi and Bertolani, 1988). In many tardigrade species, particularly echiniscids, females are unknown (thelytoky) and parthenogenesis is thought to be the normal reproductive strategy. Parthenogens have reproductive advantages over amphimictic populations. A parthenogen can colonize a habitat more rapidly, since only a single animal need invade a new territory to found a new population. In addition, sexually reproducing species have to expend energy on mate location, copulation and fertilization, thus reducing the relative returns from the reproductive process. The reproductive capacity of a parthenogen is, however, achieved at the expense of a reduction in genetic variability and, consequently, a reduced potential for adaptation (particularly valuable in changing environments). The evolution of parthenogenesis in tardigrades may, therefore, be linked to the evolution of cryptobiosis — a mechanism which effectively maintains activity within restricted environmental parameters by avoiding extremes (Pilato, 1979). This is supported by the fact that both parthenogenesis (Bertolani, 1987a) and cryptobiosis (Wright et al., 1992) are rare in marine tardigrades.

Polyploidy is often associated with parthenogenesis. Triploids are most common, but tetraploids have been observed in a few cases (Bertolani, 1973b; Rebecchi, 1991). Very slight morphological differences of the sclerified structures (particularly egg shells) observed in the diploid cytotypes and corresponding polyploids of *Ramazzottius* spp., are considered by Rebecchi and Bertolani (1988) to be indicative of the relatively recent origin of parthenogenetic forms.

Spermatozoa

Spermatogenesis has been studied in few species. *Isohypsobius granulifer* was studied by Wolburg-Buchholz and Greven (1979) and *Batillipes noerrevangi* by Kristensen (1979). A review is given by Bertolani (1983c). The process appears to start early in the animal's development, possibly after the first moult. As spermatozoa from so few species have been analysed, discussions of the phylogenetic significance of structural variation can only be speculative. There are at least three different types of tardigrade spermatozoa. A form with a single flagellum and a rounded head has been observed in the Heterotardigrada (Kristensen, 1979), while eutardigrades have forms with a helicoidal head and fringe (Baccetti *et al.*, 1971; Rebecchi and Guidi, 1991) or an undulatory membrane (Baccetti, 1987). Active sperm can sometimes be observed when males are squashed under the microscope — even from animals which have just emerged from anhydrobiosis — suggesting that fully formed gametes are unaffected by dehydration (I. M. Kinchin, unpublished work). The forms observed in the Tardigrada are considered by Baccetti (1987) to be consistent with the relatively recent acquisition of a semi-terrestrial existence, since they lack the peculiar specializations that are typical of other 'terrestrial' phyla. However, as tardigrade evolution is thought to have been very slow (Pilato, 1979), morphological adaptation of spermatozoa may have lagged behind that of other phyla.

Eggs and Embryos

Oogenesis has been described as proceeding with a 'certain haste' in tardigrades, resulting from the fact that reproduction has to be completed during periods of favourable environmental conditions which are often of short duration (Węglarska 1979a). The mechanism of formation of egg shell processes is poorly understood. The observations of Hallas (1972) suggest that the processes form as the egg surface area decreases along with a decrease in the volume of the ovum (an ovum is often larger than the final egg). The shell processes grow from the margin of their bases, and so the distal part is formed first. The number of eggs produced varies tremendously, even within a single species. Kathman and Nelson (1987) observed the number of eggs laid in exuvia in a population of *Pseudobiotus augusti* to range from 3 to 35.

Egg-shell morphology is considered to be a powerful tool in eutardigrade taxonomy. The eggs of *Macrobiotus* cf. *hufelandi* have finely ornamented shells (Figure 5.3) which have been studied in some depth (Grigarick *et al.*, 1973; Toftner *et al.*, 1975; Biserov, 1990a; 1990b; Bertolani and Rebecchi, 1993), and have been used to separate species within the *hufelandi* complex. A number of

Figure 5.3. SEMs of eutardigrade eggs
(A) *Macrobiotus areolatus*; (B) *Macrobiotus* cf. *hufelandi*. SEMs courtesy of Diane Nelson. (Bar ≈ 5 μm.)

the species within the *Ramazzottius oberhaeuseri* group are also only easily separable on the basis of egg morphology (see Chapter 9).

The complexity of tardigrade egg shells, and the amount of energy presumably invested in their production, suggests that it is an important factor in their survival. The function of egg-shell ornamentation is not known, but the following hypotheses are suggested.

- To help maintain the egg's position within the substratum, preventing dislodgement by water movements or other disturbances.
- To avoid predation by preventing penetration by nematode buccal stylets or adhesion of fungal appressoria on a smooth surface.
- To prevent rapid dehydration by trapping pockets of water between the projections when the substratum dries out.
- To provide a surface for gaseous exchange. The elaborate ornamentation of many tardigrade egg shells is structurally very similar to that forming a chorionic plastron on the surface of many terrestrial insect eggs (Hinton, 1969). In insects, the interstices between the hydrophobic struts of the chorionic meshworks hold the gas. The chorionic plastron may become filled with gas while still bathed in the fluid of the oviduct. This involves active absorption of the contained liquid which reduces the pressure in the intrachorionic spaces and causes bubbles of gas to be formed. The capacity of avoiding rapid water loss is not necessarily impaired by the provision of a plastron and so does not conflict with the tardigrades' physiological requirements for anhydrobiosis. Hinton (1969) considers such a plastron to confer greatest advantage in environments that are alternately wet and dry, such as those inhabited by bryophilous tardigrades.

However, chorionic plastrons are not universal in the animal kingdom. For example, while the morphology of oxyurid nematode egg shells suggests that they may have evolved similar adaptations to those of insect eggs, they do not appear to fulfil similar functions (Wharton, 1980). Detailed studies of the respiratory physiology of tardigrade eggs is required before the chorionic plastron hypothesis can be confirmed or rejected.

Eggs can be laid freely in the substratum or enclosed in the cast exuvium of the adult. Those laid in an exuvium often have smooth, unornamented shells. In these cases, the exuvium performs the same functions as the ornamentations on free eggs. In some cases, the exuvium full of eggs is dragged around by the female (Figure 5.4). This constitutes the only form of parental care exhibited by tardigrades. Those laid freely are sometimes laid in pairs or small groups and so, if one is found, others are often observed in the same sample. It is presumed that in stratified microhabitats, such as moss cushions, the eggs are laid in the layers which offer optimum conditions of hydration. There has been speculation that egg-laying in some species (particularly those with eyespots) may be triggered by changes in the photoperiod (Morgan, 1977; Kinchin, 1985), but

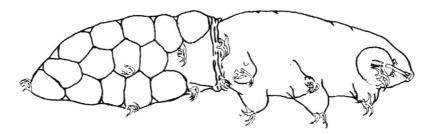

Figure 5.4. *Pseudobiotus megalonyx* female with exuvium full of eggs
Reproduced from Bertolani (1982b), with permission.

there have been no controlled laboratory observations to support this. Eggs of most marine species are unknown.

Most of the detailed studies of tardigrade embryological development date from last century and the first half of this century [Marcus 1928; 1929a, see Bertolani (1990) for a summary]. At gastrulation, the major groups within the taxon cannot be distinguished. The embryo develops a curvature which helps in determining the dorsal and ventral surfaces (Figure 5.5). The proctodaeum is formed at the posterior end of the embryo by ectodermal invagination. A second invagination at the anterior end forms the stomodaeum. Various interpretations have been placed upon observations of tardigrade embryological development. Authors disagree about the nature of the body cavity which has been described as a coelom, a pseudocoelom, a schizocoelom and a mixocoelom (Marcus, 1929b; Pollock, 1975a; Greven, 1980; Nelson, 1982a).

The posterior somatic pouch forms the gonad — the only eucoelomic organ. Body cavity cells and muscles originate from isolated cells (anlage) formed from the disaggregation of the enterocoelous pouches. The midgut is derived from the archenteron, while the hindgut and pharynx are derived from the proctodaeum and stomadaeal thickenings, respectively. The legs are formed from ventral folds in the ectoderm. The duration of embryological development varies with temperature and the persistence of suitable environmental

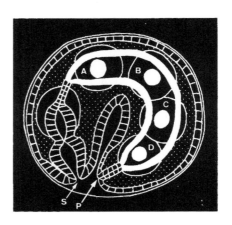

Figure 5.5. Sagittal section of an embryo of *Hypsibius convergens*
The stomadaeum (s) and the proctodaeum (p) can be seen towards the bottom of the Figure. The stomadaeal thickening is the precursor of the pharyngeal bulb. Four somatic pouches (A–D) are shown along the length of the midgut. Redrawn from Marcus (1929b).

factors (i.e. it may be interrupted by cryptobiosis). Typically, development takes from one to three weeks. Embryos in the latter stages of development are frequently visible in eggs, particularly once the pharyngeal apparatus has been formed. A complete set of buccal apparatus can often be seen in mature eggs (Figure 5.6) and this helps to confirm the species to which the egg belongs.

Hatching and post-embryonic development

It is thought that some species use their buccal stylets to pierce the egg shell and facilitate hatching of the first instar. However, this is unlikely to be the case for all species, particularly those with small, weak stylets. During hatching, and the hours immediately afterwards, the animal increases in volume by pumping fluid into the intestine using the pharyngeal bulb. This swelling of the body is probably as important as the action of the stylets in helping the animals burst out of the egg. The folded cuticle of the embryo is straightened and the final length of the first instar animal is typically about three times the diameter of the egg (Hallas, 1972). Variations in egg size will, therefore, determine the variation in size of first instar individuals (and possibly also variations in adult sizes). The eggs of some species are known to vary considerably in size. For example, in a population of *Milnesium tardigradum*, eggs were observed to range in size from 69 μm to 114 μm (Schuetz, 1987). Egg diameter sets the upper size limit of juveniles, since the inflexible portions of the buccal apparatus (stylets and buccal tube) cannot be reduced in size by folding or other contortions to fit into the egg.

Whereas the eutardigrades display a direct post-embryonic development, with the juveniles appearing as smaller versions of the adult habitus, the heterotardigrades (with a few exceptions) display an indirect post-embryonic development, with the animals reaching their definitive habitus through a succession of moults in three stages, as recognized by Bertolani *et al.* (1984). In stage 1, the legs have digits or claws numbering two fewer than in the adult form. The anus and gonopore are absent. This is sometimes described as a larval stage and is

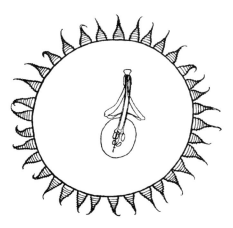

Figure 5.6. Drawing of an embryonated egg from *Macrobiotus occidentalis*
Reproduced from Dastych (1984), with permission.

thought to be of short duration. At stage 2, the adult number of digits or claws is attained, but the gonopore is absent or sometimes present in a rudimentary form. The perianal folds of cuticle are developed, but the reproductive structures are absent. By stage 3, in addition to the adult claw number, the gonopore is present and completely developed.

Some exceptions to this general scheme include *Batillipes noerrevangi*, which has at least four larval stages which are characterized by a gradual increase in the number of digits (Kristensen, 1978), and *Tetrakentron synaptae* in which, as an adaptation to the parasitic mode of life, the larval stages are suppressed (Van der Land, 1975; Kristensen, 1980). Some specimens belonging to the Stygarctidae are described as displaying neoteny in claw number — only two claws are present on each leg (usual for first stage larvae) although mature seminal vesicles are also visible (Noda, 1993).

There are other examples of morphological changes associated with development in some arthrotardigrade species, including small changes to the cephalic or caudal appendages in some *Batillipes* spp. (McGinty and Higgins, 1968; Grimaldi de Zio and D'Addabbo Gallo, 1975) and the formation of cuticular expansions in *Actinarctus* spp. (Grimaldi de Zio *et al.*, 1980; 1982c).

Development of a number of echiniscid species is characterized by changes in the cuticular texture (e.g. Lattes, 1975), or an increase in the number and size of cuticular appendages (e.g. Iharos 1961; Grigarick *et al.*, 1975; Ramazzotti and Maucci, 1983). The development of cuticular appendages in *Testechiniscus spitsbergensis* is shown in Figure 5.7. *Mopsechiniscus imberbis* represents an exception to this trend. The number of appendages decreases from the larval stage until they are absent in the adult form (Ramazzotti and Maucci, 1983). While an increase in cell size is an important component of post-embryonic growth in tardigrades, mitotic cell division also occurs so that, with the possible exception of the nervous tissue, eutely apparently does not occur in tardigrades (Bertolani, 1970a; 1970b).

Moulting

Periodic moulting continues throughout the life of all tardigrades. The process begins with the expulsion of the stylets and buccal tube (Figure 5.8.1). The mouth opening closes up so that this 'simplex' stage is unable to feed until ecdysis is complete and the buccal apparatus is regenerated. The muscles associated with the buccal apparatus stay in place and are surrounded by the salivary glands which move forward as the cuticular structures are expelled. The salivary glands control the regeneration of the buccal tube, the stylets and their supports, while the placoids are regenerated by the epithelial cells of the pharyngeal bulb (Figure 5.8.2–5.8.5).

The claws are reformed during each moult. These are formed by an organ at the distal end of each leg — the claw gland (Figure 5.9). Kristensen (1976) has described in detail the process of claw formation in the heterotardigrade

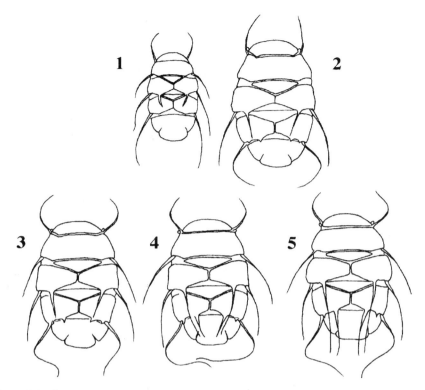

Figure 5.7. Developmental stages of *Testechiniscus spitsbergensis*
(1) Two-clawed juvenile; (2) four-clawed juvenile; (3 and 4) adult stages; and (5) typical adult form. Redrawn from Iharos (1961).

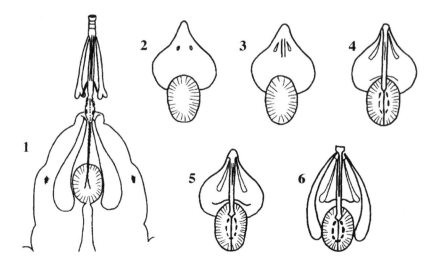

Figure 5.8. Drawing of expulsion of pharyngeal apparatus (1) and stages in the regeneration of the stylets and placoids (2–6)
Redrawn from Marcus (1929b).

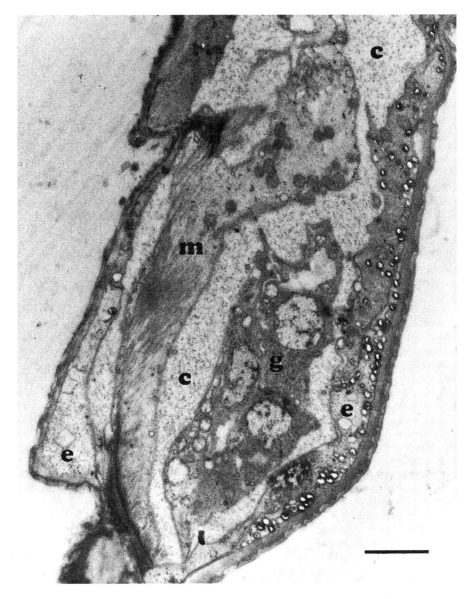

Figure 5.9. TEM of a section of the distal portion of a eutardigrade leg (*Ramazzottius varieornatus*) showing the claw gland
Abbreviations used: c, body cavity; e, epidermis; g, claw gland; m, muscle cell. Bar = 2 μm.

Batillipes noerrevangi. The distal pole of the claw gland forms deep infoldings from which emerge the new claws, formed by the secretion of cuticular materials by the gland. The process in eutardigrades is identical (Walz, 1982). In many tardigrades, moulting is related to vitellogenesis, particularly when eggs are laid in the exuvium.

When observing a population of tardigrades, a number of simplex forms may be seen in which the new claws are visible under the 'old' cuticle, prior to ecdysis. This can be seen in eutardigrades, but is most obvious in the large claws of some marine heterotardigrades such as *Echiniscoides sigismundi* (e.g. Grohé, 1976) and *Batillipes* spp. (Figure 5.10). As juvenile heterotardigrades lack an anus, the first moult is the first opportunity to expel waste from the intestine. In *Batillipes mirus*, the second instar defecates into the old cuticle during the first ecdysis (Figure 5.11).

Processes such as defecation, oviposition and ecdysis are associated with changes in the pressure of the fluid in the body cavity. When the fluid pressure is reduced, after defecation or oviposition, moulting is made easier as the new/inner cuticle is pulled away from the old/outer cuticle. Consequently, these processes are often syncronized. Internal fertilization is also easier during periods of reduced fluid pressure (Pollock, 1975a).

Higgins (1959) used plots of body length frequency to determine the number of instars in a population of *Macrobiotus islandicus*. Because tardigrades periodically shed their cuticle to grow, a correlation can be made between the moults and intermoults and the troughs and peaks of the graph. This technique has subsequently been used by other authors (e.g. Franceschi Crippa and Lattes, 1967; Morgan, 1977; Wainberg and Hummon, 1981) to determine the number of instars in eutardigrade species, with estimates ranging from 6 to 13. However, few authors have been able to show the clear transitions

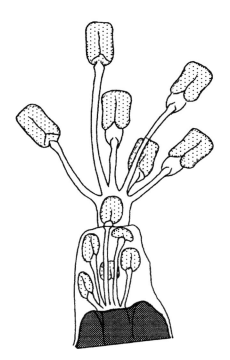

Figure 5.10. Regeneration of *Batillipes* digits
Redrawn from Marcus (1929b).

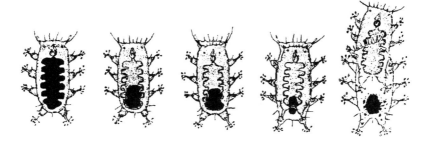

Figure 5.11. First moult in *Batillipes mirus*
The second instar is the first to have a functional anus and so defecates at the earliest opportunity — upon emergence into the shed cuticle. Reproduced from Marcus (1929b,) with permission.

from one instar to the next that might be expected (Figure 5.12). The technique is faced with the following problems

- If a population has an annual cycle of development, measurements gained may vary if the population is sampled at different times of the year as it is possible that not all instars will be represented in a single sample — particularly in univoltine populations. This has been demonstrated for *Macrobiotus hufelandi* by Franceschi *et al.* (1962–1963), *Macrobiotus* (now *Dactylobiotus*) *grandipes* by Schuster *et al.* (1977) and for *Pseudobiotus augusti* by Kathman and Nelson (1987). To overcome this problem, the population should be sampled regularly throughout the year. Data from single samples, therefore, cannot be relied upon.
- It is possible that the population consists of more than one physiological race or 'cytotype' [members of the same species are morphologically identical but exhibit different chromosome numbers, as demonstrated for *Macrobiotus hufelandi* by Bertolani (1982a) and for *Macrobiotus pseudohufelandi* by Rebecchi (1991)] which may possibly exhibit varying numbers of instars. This may help to explain the range of estimates of instar numbers, from 9 to 12 for *Macrobiotus hufelandi* (Marcus, 1929b; Franceschi Crippa and Lattes, 1967; Morgan, 1977) and from 6 to 12 for *Macrobiotus areolatus* Ramazzotti (1977). These differences may, alternatively, indicate that a number of similar species are being 'recognized' as a single species (see Chapter 9; *Macrobiotus* cf. *hufelandi*)
- Males and females are known to be of differing sizes in some species. It is, therefore, necessary to measure males and females separately. Unfortunately, it is not often possible to separate the sexes as there is little observable sexual dimorphism (Nelson, 1982a; Bertolani, 1992), particularly in the immature specimens representing the first two or three instars. A bimodality in size frequency graphs was noted by Kathman and Nelson (1987) for a population of *Pseudobiotus augusti* and was attributed to the presence of separate sexes, though these could not be separated morphologically.

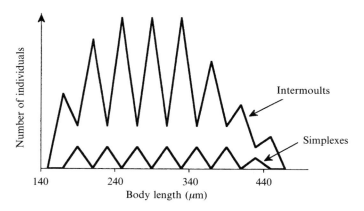

Figure 5.12. Graph showing an idealized plot of the size frequencies in a tardigrade population of a hypothetical species exhibiting eight instars
The intermoult peaks alternate with the simplex peaks.

- Hallas (1972) has discussed the reasons why the body length of first instars should not deviate significantly from three times the diameter of the egg. Variation in egg diameter will, therefore, lead to variation in the size of juveniles. In *Dactylobiotus grandipes* (Schuster *et al.*, 1977) the first instar has been calculated to vary in length by 123 μm (from 177 μm to 300 μm) on the basis of variation in the diameter of eggs. Such a degree of variation in each of the instars would account for considerable blurring of the transition from one instar to the next. Morgan (1977) has also suggested that instars may develop at varying rates with juveniles developing more rapidly than mature individuals — explaining the small number of juveniles typically observed.

Cyclomorphosis

The marine tardigrade *Halobiotus crispae* has been observed to undergo an annual cycle of morphological change (cyclomorphosis), alternating between two distinct morphs (Kristensen, 1982*a*). This is similar to that observed in the Collembola and is characterized by changes within the individual rather than changes in successive generations as observed in some rotifers. The summer morph is sexually mature whereas the sexually immature winter morph is a specialized hibernating stage which is highly tolerant to freezing temperatures and low salinities (Figure 5.13).

It is not clear how widely the phenomenon of cyclomorphosis is spread among the tardigrades. Observations of simplex and pseudosimplex stages may have led to some false identifications in the past. For example, specimens described as *Itaquascon umbellinae* (De Barros, 1936) are now regarded as the simplex stage of *Diphascon* (now *Mesocrista*) *spitzbergense* (Kristensen, 1982*a*). However, doubts about the validity of the entire genus *Itaquascon*, and that

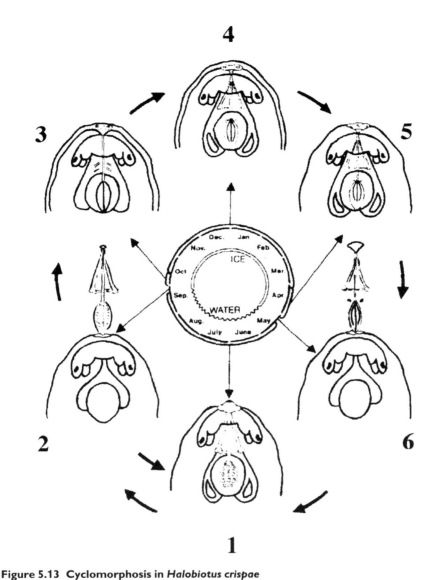

Figure 5.13 Cyclomorphosis in *Halobiotus crispae*
(1) Active, mature; (2) simplex; (3) prae-pseudosimplex; (4) pseudosimplex 1; (5) pseudosimplex 2; and (6) simplex. Reproduced from Kristensen (1982a), with permission.

specimens simply represent the simplex stages of various *Diphascon* species, have been quashed following observations of sexually mature individuals and studies of their chromosomes (Bertolani, 1979a; Kristensen, 1982a).

6

Cryptobiosis

The need for latent states

The marine Arthrotardigrada (the most primitive order of tardigrades) do not exhibit the ability to suspend metabolism using any of the processes collectively known as cryptobiosis[1]. The marine environment is relatively constant with no rapid fluctuations in temperature or humidity and so physiological mechanisms to withstand such environmental perturbations are unnecessary. However, because these marine organisms invaded limno-terrestrial environments some time ago in the geological past, a requirement to tolerate drying became vital, since the small bodies of freshwater typically colonized by tardigrades are less stable than the marine environment. May (1953) has suggested an evolutionary progression from the marine arthrotardigrades to the 'terrestrial' echiniscids and eutardigrades. He cites a number of intermediary stages which display limited tolerance to environmental fluctuations. For example, *Echiniscoides sigismundi* living amongst *Enteromorpha* lawns and barnacle plates high in the littoral zone, has to cope with regular short-term desiccation between tides, accompanied by fluctuations in oxygen tension. Additionally, heavy rains at low tide will cause a great change in the water potential of the surrounding medium. *Echiniscoides sigismundi* survives these unfavourable periods in an immobile state, resuming activity when covered again by the tide.

May (1953) concluded that the non-armoured Echiniscoidae point the way to the history of the terrestrial invasion, from whose ancestors the Echiniscidae and Eutardigrada have evolved. It is only these two groups which exhibit the typical characteristics of cryptobiosis described below. The evolution of cryptobiosis was a prerequisite to the successful invasion of terrestrial micro-

---

[1] While marine species seem generally incapable of survivng periods of cryptobiosis, Higgins (1977a) reported that (unnamed) tardigrades could be revived from beach sand that was stored for several months in a dry condition.

environments by tardigrades and other aquatic meiofaunal groups (e.g. nematodes, rotifers), and represents a point of convergence between these taxa which has had profound implications upon their ecology and continued evolution (Pilato, 1979). By avoiding unfavourable conditions in a dormant/resistant state, cryptobionts are effectively screened from environmental selection pressures resulting in a slow rate of evolutionary change (bradytely). They are, however, able to colonize habitats in which most aquatic microfaunal groups would not be able to survive and so avoid competition.

The term 'cryptobiosis' was coined by Keilin (1959) to replace the terms 'abiosis' and 'anabiosis', describing a reversible suspension of metabolism initiated by adverse changes in one or more environmental factors. This is distinguished from 'quiescence', which is only the slowing down of metabolism under unfavourable conditions and from 'diapause' which is a hormonal suppression of development independent of short-term environmental cues[2] (Womersley, 1981). Significant metabolic activity is detectable in encysted tardigrades (Pigoń and Węglarska, 1953), and so this may be described as a form of quiescence distinct from cryptobiosis.

Latent states are quite common in the Animal and Plant Kingdoms. Typically, they occur at certain stages in the life-cycle of the organism, particularly the propagules (e.g. the seeds of various flowering plants, the resistant eggs of cestodes and so on). In tardigrades, however, latent states can seemingly occur at any stage of the life-cycle. Cryptobiosis in tardigrades has been reviewed by Crowe (1975) and Wright *et al.* (1992) and has been divided, according to the dominant influencing environmental factor, into anhydrobiosis (dehydration), cryobiosis (temperature), osmobiosis (water potential) and anoxybiosis (oxygen tension). Cryptobiosis can extend the life-span of a tardigrade from a few months up to several years (possibly over a century), although the active life-span is unaffected.

Anhydrobiosis

The phenomenon of anhydrobiosis was noted by the earliest observers of tardigrades when animals were seen to revive from dehydrated sediments. In the 18th Century, Spallanzani described tardigrades as a group which could enjoy the advantages of "resurrection after death". While the terminology may have changed over the past two hundred years, the fascination created by anhydrobiosis persists and the apparent paradox of life without water has been eloquently summarized by Crowe (1975):

---

[2] *Dormancy in arthropods has been described as either 'predictive' or 'consequential' by Müller (1970). Predictive dormancy (diapause) is initiated in advaance of the adverse conditions and is typically found in predictable or seasonal environments. Consequential dormancy (including cryptobiosis) is initiated by a change in environmental conditions and is typically found in unpredictable environments.*

"If metabolism actually comes to a halt in anhydrobiotic tardigrades, we are forced into a seemingly insoluble dilemma about the nature of life; if life is defined in terms of metabolism, anhydrobiotic tardigrades must be 'dead', returning to 'life' when appropriate conditions are restored. But we know that some of the organisms die while in the anhydrobiotic state, in the sense that they do not revive when conditions appropriate for life are restored. Does this then mean that they 'died' while they were 'dead' ?"

Anhydrobiosis is arguably the most important of the types of cryptobiosis and has probably been the most extensively studied. The factors affecting anhydrobiosis in tardigrades can be divided as follows.

- Extrinsic factors
  — Dehydration of the surrounding medium.
  — Environmental encapsulation.
  — Behavioural responses.
- Intrinsic factors
  — Tun formation.
  — Cuticular permeability.
  — Membrane protectants.
  — Energy reserves and recovery.

Microhabitats of various types will dehydrate at different rates under similar physical conditions. The factors affecting dehydration rates of moss cushions are discussed in Chapter 7. Some microhabitats (e.g. soil) may dry unevenly to result in the encapsulation of animals in a pocket of moisture in a drying medium. Alternatively, the microhabitat may react to desiccation (e.g. moss leaves curling up around the stem), again slowing the rate of water loss from the animal's immediate microenvironment.

Aggregation behaviours, similar to those observed in anhydrobiotic nematodes (e.g. Croll, 1970), have been observed in tardigrades inhabiting xeric environments (e.g. Kinchin, 1989a), although no quantitative data are available. Clumping behaviours reduce the total surface area of the group and so reduce the overall rate of water loss. This is obviously more beneficial for those animals in the middle of the clump than those on its edges. Damp debris may also be incorporated into these clumps and help to retard water loss still further. Alternatively, tardigrades may migrate within the microhabitat in order to maintain their position in a region of optimum hydration (Wright, 1991).

Formation of a tun seems to be vital for tardigrades to achieve anhydrobiotic success, and a slow rate of dehydration is, in turn, essential for successful tun formation (Crowe, 1972). The bodies of animals which are dried too quickly simply collapse and do not revive upon rehydration. The tun is formed by invagination of the limbs, longitudinal contraction of the body and infolding of the intersegmental cuticle (Figure 6.1). A reduction in exposed

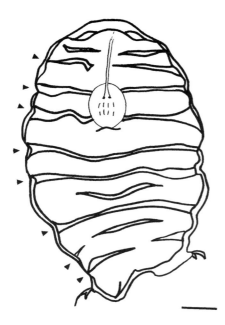

Figure 6.1. Tun of *Ramazzottius varieornatus* (dorsal view)
Arrows indicate infolding of intersegmental cuticle. Bar = 30 μm. Reproduced from Kinchin (1990c), with permission.

surface area seems to be a universal adaptation in anhydrobionts —nematodes coil their bodies to achieve this (e.g. Demeure *et al.*, 1979), while rotifers undergo considerable antero–posterior contraction (e.g. Ricci and Melone, 1984).

A rapid decline in cuticle permeability has been observed to occur soon after completion of tun formation (Crowe, 1972; Wright, 1989a). This phenomenon is described as a permeability slump and may reduce cuticular transpiration by a factor of 20–60. This is observed in both living and dead animals and is, therefore, not a consequence of metabolic processes (Wright, 1989a). Eutardigrades show a positive correlation between surface area reduction in the tun and desiccation tolerance. In addition, the permeability slump is effected more rapidly in more tolerant species. These species also display the thickest intrasegmental cuticle and the greatest degree of thinning in the intersegmental cuticle. The earlier initiation of the permeability slump in more tolerant species would allow them to withstand more rapid transpiration and, hence, lower humidities. This is what is found (Wright, 1989a). Xerophilous species such as *Ramazzottius* cf. *oberhaeuseri* tolerate humidities down to 59%, while hygrophilous species, such as *Hypsibius dujardini*, will only tolerate a minimum of 78%. In addition, some tolerant eutardigrades with pigmented epidermis lack pigment under the intersegmental cuticle, resulting in characteristic dorsal stripes of the *oberhaeuseri*-group, for example. These unpigmented bands may help to confer greater flexibility to these areas, facilitating rapid and effective tun formation. Lipids have been observed to be extruded from pores in the cuticle, and may help to reduce transpiration and/or provide protection against fungal attack (Wright, 1988b).

In terrestrial heterotardigrades, the epicuticle is modified as thickened dorsal plates. These rigid structures restrict surface area reduction during tun formation. The thinner intersegmental cuticle folds to allow the edge of the plates to slide over each other to retard water loss from the dorsal surface. In addition, the air-filled lacunar system of the heterotardigrade cuticle (see Chapter 4) may also help to reduce transpiration by creating an extended diffusion pathway (Wright, 1989*a*). Most of the reduction in surface area of anhydrobiotic echiniscids is achieved by invagination of the limbs and folding of the unarmoured ventral surface (Figure 6.2).

Anhydrobionts achieve a degree of dehydration that not only involves the loss of 'free water', which forms the aqueous solutions of the body, but also the loss of 'bound water', which is required to maintain the structure of vital hydrated macromolecules such as proteins, membrane phospholipids and nucleic acids. One of the problems facing anhydrobionts is that of maintaining the structural integrity of their tissues, while simultaneously removing all their water. Destruction of these hydrated macromolecules, by loss of bound water, would cause irreversible damage to the animal and result in anhydrobiotic

Figure 6.2. Ventral view of a tun of *Echiniscus* sp. (Heterotardigrada)
Bar = 10 μm. SEM by R.O Schuster, courtesy of Diane Nelson.

failure. The bound water, therefore, has to be replaced by a compound which will maintain structural integrity during dehydration and which is easily removed during rehydration. Trehalose and glycerol have been identified as membrane protectants in anhydrobiotic arthropods and nematodes (Clegg, 1964; 1965; Madin and Crowe, 1975). Trehalose is produced in direct response to dehydration (Womersley and Smith, 1981; Westh and Ramløv, 1991) (Figure 6.3). Trehalose is a non-reducing sugar and so will not participate in browning reactions between sugars and amino groups (Crowe and Clegg, 1973) and will actually inhibit browning between dry proteins and reducing carbohydrates present in anhydrobiotic organisms (Loomis et al., 1979), so aiding the animal through the period of dehydration[3].

Under these conditions, it seems highly improbable that any metabolic activity is maintained. In addition, dried tardigrades are able to survive conditions which are beyond the extremes under which active life has been observed (Scheie, 1970), and beyond those which are normally encountered in nature, such as exposure to vacuums, X-rays, u.v. radiation and temperatures approaching absolute zero[4] (Ramazzotti and Maucci, 1983). Although metabolism has been detected in tardigrade tuns at high humidities (Pigoń and Węglarska, 1955a), where tuns are maintained in dry conditions, it seems probable that there is a total cessation of metabolic activity (Pigoń and Węglarska, 1953; 1955b; Clegg, 1973). This is coupled with a much increased resistance to environmental extremes and a suspension of ageing.

The use of membrane protectants helps to preserve the basic structure of cell organelles during anhydrobiosis (Crowe, 1975; Walz, 1979a; Wright, 1988a) so that metabolism can be resumed as soon as the animal is sufficiently rehydrated. In a study of the revival of *Macrobiotus areolatus* from anhydrobiosis, Crowe and Higgins (1967) found that revival time (the time between the restoration of an aqueous environment and the first signs of tardigrade locomotor activity) is directly proportional to the duration of the dormant state. Animals kept dry for a few days typically revived after 10–15 min. This time

---

[3] *Recent work on the role of trehalose in protecting biological systems during desiccation, has suggested that the synthesis of trehalose is not merely a necessary adaptation, but may be sufficient adaptation to preserve various macromolecules and tissues (e.g. Roser, 1991a; 1991b; Roser and Colaço, 1993).*

[4] *The unusual ability of tardigrades to withstand such extremes has been cited as evidence of a divine creation and described as 'over-adaptation' by Vetter (1990). In fact, resistance to these ultra-extremes is a direct (and, perhaps, inevitable) consequence of the development of cryptobiosis. For example, once an animal has developed resistance to the low temperatures experienced in polar regions (by arresting metabolism and replacing water with cryoprotective organic molecules), survival at lower temperatures requires no further adaptation. The fact that the ability to survive periods at absolute zero has no adaptive significance is irrelevant. Therefore, claims that over-adaptation in tardigrades could indicate that they are extra-terrestrial invaders of the terrestrial biosphere, should be taken with a pinch of salt.*

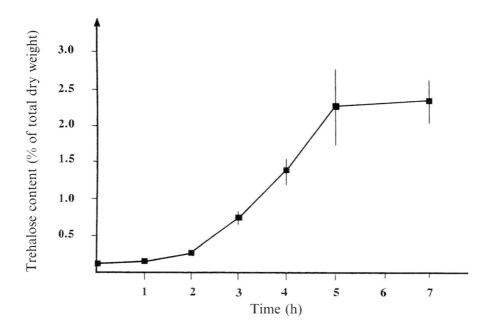

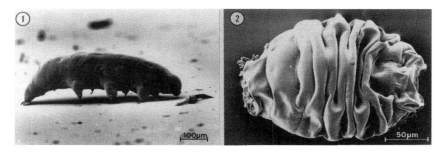

Figure 6.3. Graph showing the increase in trehalose content in *Richtersius coronifer* during dehydration expressed as percentage of total dry weight
Tun formation was completed after 3 h. Redrawn from Westh and Ramløv (1991). SEMs showing *Richtersius coronifer* in (1) the active state and (2) the tun state. SEMs courtesy of Reinhardt Kristensen.

probably varies from species to species, and is also affected by temperature, pH and dissolved oxygen.

Effects of nutrition on anhydrobiotic success have been recognized by a number of observers. Pigoń and Węglarska (1955*b*) attempted to standardize the nutritional condition of their specimens by starving them prior to treatment, while Crowe and Higgins (1967) attempted to use well-fed animals by keeping them in a moist chamber containing moss for 24 h before treatment. Observations on tardigrades and nematodes (in which the process of anhydrobiosis appears to be very similar) suggest that anhydrobiotic success is directly related to the lipid reserves of the animal (Demeure et al., 1978; Preston and Bird, 1987; I.M. Kinchin, unpublished work). This may help to explain why repeated dehydration and rehydration (with its associated effects on lipid reserves) leads to reduced viability (Crowe, 1975; Perry, 1977).

Lipids may have one or a number of the following functions in anhydrobionts.

- Large lipid droplets (particularly those in body cavity cells) may help to maintain spatial distribution of the tissues within the shrunken body and prevent reactions between molecules that are usually separated in the active animal (Bird and Buttrose, 1974).
- Lipids may be converted to glycerol or trehalose (Madin and Crowe, 1975; Westh and Ramløv, 1991) which act as membrane protectants (Crowe et al., 1984; 1987) helping to maintain the cellular integrity of the dried animal.
- The lipid may act as an energy reserve for metabolic preparations at the induction of anhydrobiosis or for use upon rehydration (Womersley, 1981).
- Cuticular lipids seem to have a role in modifying the permeability of the cuticle to water during dehydration (Wright, 1988b; 1989a; Wharton et al., 1988).

Cryobiosis

Limno-terrestrial tardigrades are common in polar regions where water is frozen for much of the year (e.g. Everitt, 1981; Kristensen, 1982a). To colonize such areas, the animals have to maintain viability in conditions in which activity is not possible. These animals are unable to avoid freezing and so must be able to cope with the physiological changes that accompany freezing. The combination of adaptive mechanisms that allow certain animals to cope with extracellular ice formation is described as 'freeze-tolerance' (Lee, 1989). Cryobiosis is an extreme form of freeze-tolerance characterized by sustained viability following low-temperature-induced metabolic arrest. To survive freezing, animals must exhibit biochemical adaptations to satisfy the following conditions (Storey and Storey, 1990).

- Controlled ice growth must be initiated in extracellular fluids. Ice nucleating agents (INAs) — probably proteins with a molecular mass in excess of 200 kDa — are added to these fluids to ensure that initial freezing processes result in the dispersal of numerous small ice crystals throughout the extracellular spaces (Westh and Kristensen, 1992; Westh et al., 1991). While INAs seed ice-crystal formation, antifreeze proteins may be used to stabilize the crystals at a small size to prevent cellular damage.
- Cell structure and function must be protected. As extracellular ice is formed, the extracellular fluid becomes more concentrated, causing water to flow out of the cells and reduce the cell volume. If the volume is reduced

below a critical minimum, the phospholipid bilayer of the cell membranes becomes compressed and its structure breaks down. Membrane protectants, such as the disaccharide trehalose, are used to stabilize membrane structure (Rudolph and Crowe, 1985). Key enzymes involved in the synthesis of cryoprotectants respond uniquely to low temperatures (increasing activity), while other enzymes are inactivated. This results in a redirection of the flow of carbon from the normal routes of carbohydrate catabolism to pathways leading to cryoprotectant synthesis (Storey and Storey, 1990).

While anhydrobiosis and cryobiosis have generally been considered to be similar processes (e.g. Lee, 1989), since both drying and freezing involve removal of water from cells, Crowe *et al.*, (1990) have suggested that the observed similarities between these processes are only superficial. For example, about 25% of the water in a cell is 'non-freezable' and so is not removed by freezing, whereas in anhydrobiosis the non-freezable water is removed by desiccation. However, the adaptation to anhydrobiosis and total desiccation is considered to be advantageous in tardigrades, such as *Richtersius coronifer*, for survival during the less extensive dehydration induced by freezing (Ramløv and Westh, 1992).

Osmobiosis

Some tardigrade species are clearly able to tolerate variations in salinity, including those found in the littoral zone (such as *Archechiniscus marci* and *Echiniscoides Sigismundi* as mentioned on page 75) and certain euryhaline limno-terrestrial species which have been observed occasionally in marine environments — *Ramazzottius oberhaeuseri*, *Milnesium tardigradum* and *Echiniscus quadrispinosus* (Renaud-Debyser, 1964; Morgan, 1976). Bryophilous tardigrades contract rapidly into a tun upon immersion in saline solutions, suggesting that the primary response is to dehydration (Wright *et al.*, 1992). However, unlike anhydrobiosis, osmobiotic survival does not require tun formation (Collin and May, 1950). Viability decreases with prolonged immersion in even moderate salinities and higher salinities are lethal within a few minutes. LD_{50} mortality estimates for *Ramazzottius* cf. *oberhaeuseri* show a survival of over 11.5 h in 2% $NaCl_{(aq.)}$ falling to 30 min in 10% $NaCl_{(aq.)}$ (Wright *et al.*, 1992). This is, however, supplemented by a significant capacity for acclimation. When returned to distilled water, osmobiotic tardigrades remain contracted in a tun for variable periods before resuming activity with revival times increasing exponentially following increasingly severe treatments (Wright, 1987). There is no evidence for the synthesis of membrane protectants in osmobiotic tardigrades.

Anoxybiosis

Given the presumed sequence of events in tardigrade evolution (i.e. a gradual transition from the marine habit to the 'terrestrial' habit) as described by May

(1953), it seems probable that a mechanism to tolerate low oxygen tensions was developed at an earlier point in tardigrade evolution than the tolerance of dehydration, since the former is more likely to be encountered in marine environments than is the latter. For example, it has been shown that nematodes and arthropods inhabiting the surfaces of the brown algae found on rocky shores, are subject to periods of little or no oxygen when the structure of these plants collapses at low tide (Wieser and Kanwisher, 1959). The animals are then covered by a dense mat of impermeable mucilaginous material until the tide rises again to oxygenate the interstices of the algal thalli. Regular variations in oxygen tensions of this type are also experienced by littoral tardigrades, such as *Echiniscoides sigismundi*, in carpets of the green alga, *Enteromorpha*. In addition, Kristensen and Hallas (1980) reported the survival of *Echiniscoides* spp. in sealed vials for 6 months, although the adaptive significance of such long-term tolerance is unknown; a viability of only 3–4 days is more typical for bryophilous species (Crowe, 1975). The observed paralysis of these animals in anoxic environments may be caused by the accumulation of toxic metabolites within the body (von Brand, 1946). The gradual desiccation of a moss cushion will also cause a reduction in oxygen tension in the interstitial water columns of a moss cushion, and so bryophilous species will also be subjected to variations in oxygen tensions.

As a result of their small size and aquatic habit, tardigrades need no specialist gas-exchange structures, but rely upon diffusion across the body wall. The oxygen consumption of tardigrades has been calculated by Pigoń and Węglarska (1955b), Jennings (1975) and Klekowski and Opalinski (1989). This varies (at 6°C) from 0.055×10^{-3} mm^3/1×10^{-6} g per h in *Doryphoribius smreczynskii* to 0.10×10^{-3} mm^3/1×10^{-6} g per h in *Macrobiotus harmsworthi*.

The animals are very sensitive to changes in oxygen tension and when the environment is unable to provide for their oxygen demands, they undergo osmoregulatory failure. Therefore, unlike other forms of cryptobiosis, water uptake rather than desiccation occurs and the animals become turgid. Anoxybiosis is sometimes deliberately induced in tardigrades by microscopists prior to making microscope preparations. The turgid, extended condition helps to reduce the distortion caused by some mounting media.

Encystment

Many species of tardigrade that are prevalent in freshwater (and some bryophilous species) display the ability to form resistant cysts. Metabolism in cysts is much less than that observed in active animals, but greater than that observed in tuns (Pigoń and Węglarska, 1953). Early observations on cysts suggested than the tardigrade body degenerated into an amorphous mass in the cyst, but detailed studies by Węglarska (1957) showed that such histolysis did not occur during encystment in *Dactylobiotus dispar*. The buccal apparatus is often recognizable in cysts (Figure 6.4).

Figure 6.4. A typical freshwater eutardigrade cyst, *Isohypsibius longiunguis*
The animal has shrunk away from the outer cuticle during ecdysis so that the regions at each pole are simply areas of collapsed outer cuticle. The legs are retracted in the inner cuticle so that the only recognizable structure is the pharyngeal apparatus.

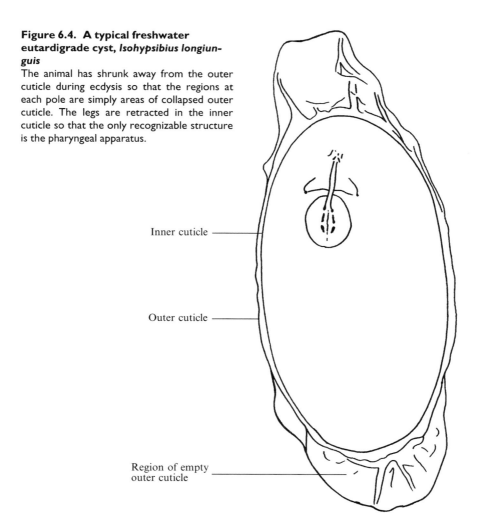

Food reserves are accumulated in the body cavity cells prior to encystment. These reserves maintain the low rate of metabolism during the period of encystment. Lack of food in the environment is thus ruled out as a factor leading to the formation of cysts. The animal enters the simplex stage upon encystment, the contents of the midgut are expelled by defecation and the cyst wall is formed by the shed cuticle. This ecdysis, therefore, differs from the normal periodic moulting in that the old cuticle is not discarded. The cuticle becomes more darkly coloured with time. The claws of the old cuticle can often be observed on the surface of the cyst.

Dimensions of the cyst tend to be much smaller than that of the active tardigrade, varying from 20–50% of the length of the active animal. Unlike tuns, cysts maintain a significant water content and, as a consequence, are a much less resistant stage. Węglarska (1957) considered that encystation is prompted by a gradual deterioration of environmental conditions. (A rapid deterioration of conditions will cause the animal to enter a state of asphyxia; animals in this condition will die after a few days unless conditions are restored.) Excystation occurs between 6 and 48 h after improvement of the environmental conditions.

These various adaptations to life in extreme, fluctuating environments have allowed tardigrades to invade habitats in which few other animals can survive. In this way, interspecific competition and many environmental selection pressures are largely avoided.

7

Ecology

Limno-terrestrial habitats

Tardigrades have most commonly been collected from moss cushions and many are considered to be bryophilous species. The characteristic of mosses which makes them such a good habitat for tardigrades is their ability to hold water within the interstices of the cushion and in the leaf axils forming bryotelmata. The cushions are, therefore, effectively isolated, temporary bodies of water that can support an aquatic fauna. The temporary nature of this habitat confers an advantage on those animals which can tolerate desiccation by using physiological mechanisms, such as anhydrobiosis, over those animals which cannot. Mosses at the dry end of the spectrum will be inhabited by a more specialized bryofauna than mosses which are usually wet.

With the exception of coastal dune mosses, in which only particular species of eutardigrade are present (Bertolani, 1983*a*), there appears to be no direct correlation between particular moss species and the species of tardigrade they support (e.g. Nelson, 1975; Kathman and Cross, 1991), but the moisture regime of a moss cushion is important in determining the composition of the aquatic fauna — including tardigrades. Mosses have been categorized according to their moisture regime (Mihelčič, 1954/55):

- Section A (wet mosses)

Group 1: submerged or floating mosses which are permanently soaked in water.
Group 2: those on the banks of waterways, or in the immediate vicinity of waterfalls, or frequently submerged by floods and which never dehydrate completely.
- Section B (moist mosses)

Group 3: those growing in shady places or humid atmospheres away from direct sunlight; these may desiccate from time to time.

- Section C (dry mosses)

Group 4: those that are exposed to direct sunlight daily and which frequently dry out.

Group 5: those that are exposed to direct sunlight for prolonged periods. These mosses are often on dry substrata, such as rocks and walls. They may dry out quickly and remain dry for long periods (periods of dehydration may be longer than periods of hydration).

Mosses in groups B and C share a drought-tolerant pattern of adaptation to dehydration with animal groups including tardigrades, nematodes and rotifers, termed poikilohydry. This adaptation is advantageous to both the bryophyte and the desiccation-tolerant aquatic bryofauna. Anhydrobiotic tardigrades are highly resistant to environmental extremes (see Chapter 6) and mosses are also more resistant when dehydrated (Nörr, 1974), so, in this way, the moss and its resident fauna can survive adverse conditions in a dormant state.

It is the moss's ability to absorb and store water which makes it the preferred microhabitat for many tardigrade species. The growth form of a moss is an important factor for the aquatic bryofauna, as it affects the moss's capacity for water storage, the distance which water must be conducted to the evaporating surface and the rate of water loss to the air. The average maximum water content of ectohydric mosses, in which tardigrades are so common, ranges from 400% to 1200% of air-dry weight . The typical rate of water loss by a moss cushion is probably a more important factor in determining the composition of the bryofauna than the typical length of dry periods, and this is reflected in the five categories listed above (Ramazzotti and Maucci, 1983). A number of factors contribute to this desiccation rate.

- The exposure of the moss to sunlight and wind.
- The nature of the substratum. Those on rock will dry more quickly than those growing directly on soil.
- The depth of the rhizoid layer and the nature of the enclosed particulate matter (particle size, humus content and so on).
- The habit of the moss. Dense cushions of acrocarpous mosses tend to dry more slowly than thin mats of pleurocarpous mosses on similar substrata.
- The outline of the cushion. Those with an irregular outline dry more quickly than those with a smooth outline, since irregularities disrupt the boundary layer of air above the plants.

Many moss species display particular adaptations to retard water loss. For example, when the cushion dries, the leaf laminae may curl up and the whole leaf may become tightly pressed against the stem. In *Tortula princeps*, the leaves wind helically around the stem so that the drying leaf is then protected from rapid desiccation by the crown of bristling hair points. This kind of leaf movement is a particularly marked characteristic of mosses from arid environments (e.g. Scott, 1982). Slowing of water loss by the moss cushion is essential to the tardigrades (and other aquatic bryofauna) living within it, since a slow rate of desiccation increases their chances of anhydrobiotic survival (see Chapter 6).

It may be speculated that as increasingly desiccation-tolerant bryophyte species evolved and invaded the dryer parts of land masses, they were accompanied by increasingly xerophilous tardigrade species. Poikilohydry (in mosses) and anhydrobiosis (in tardigrades) would then have evolved in parallel. In this way, the evolution of bryophytes may well have influenced the evolution of tardigrades. The physiological ecology of mosses (an important factor influencing the survival of bryophilous tardigrades) has been reviewed by Proctor (1979; 1982; 1984). Other cryptogams, such as liverworts and lichens, support a similar microfauna to that found in the mosses. The liverworts tend towards the wetter end of the spectrum (moss groups 2 and 3), while the lichens tend towards the drier end (moss groups 4 and 5).

A number of phanerogams produce structures which are capable of holding water and which, in turn, support a complex aquatic microfauna. Such structures (rosettes of bracts, leaf axils and so on.) are known collectively as phytotelmata. These are often observed in the tropics, the most well-known being the bromeliads and pitcher-plants (e.g. Frank and Lounibos, 1983). Such structures also occur in temperate regions, though usually on a smaller scale than in the tropics. Tardigrades have been recovered from the water collected in the rosettes of several alpine plants, such as Saxifrage and Androsace (Ramazzotti and Maucci, 1983). On a larger scale, the leaf axils of teasel plants (*Dipsacus* spp.) can hold several cubic centimetres of water, in which will be found a variety of invertebrates. Midge larvae are particularly common from this microhabitat, but tardigrades have also been collected by Maguire (1959) and Masters (1967).

Tardigrades are also common in soil and leaf litter, although estimates of population densities vary widely. Tardigrades are not evenly distributed in soils, but are typically found in dense aggregations in one area — descibed as 'nests' by Mihelčič (1963) — while absent in adjacent areas. They are often found around plant roots and are always most common in the top few centimetres of soil. There have been very few observations of soil tardigrades, but there is a suggestion that heterotardigrades are rare or absent from this habitat (Fleeger and Hummon, 1975; Manicardi and Bertolani, 1987). The species composition in soil appears to be very different from that found in adjacent moss cushions, as has been recorded for nematodes (Overgaard-Nielsen, 1948).

Other (true) freshwater habitats, such as lakes and rivers, have been sampled for tardigrades. In a survey of freshwater tardigrades from Korea, Moon *et al.* (1989) took samples from the water column of lakes and rivers with a plankton net. However, the majority of freshwater studies refer to observations form benthic or shore-line sandy sediments (e.g. Pennak, 1940; Neel, 1948; Schuster *et al.*, 1977; Wainberg and Hummon, 1981; Evans, 1982; Čuček, 1985; Kathman and Nelson, 1987; Van Rompu and De Smet, 1988; 1991), although some studies, particularly those from Antarctic regions, have concentrated on benthic algal mats as a substratum (e.g. Cathey *et al.*, 1981; Everitt, 1981; McInnes and Ellis-Evans, 1990). In general, tardigrades from the freshwater psammon are most abundant in the top few centimetres of sand. However, this trend is reversed in the wave-impacted zone on the lake shore; possibly because the animals burrow more deeply to avoid injury from sand-grain movements, or because this zone has less detrital food material than areas above and below the waterline (Pennak, 1940; Evans, 1982). In a study of Lake Cerknica in Slovenia, Čuček (1985) observed a temporal variation in tardigrade species. The hygrophilous species, which were dominant at the time of normal water levels, 'disappeared' at the regular drying of the lake. At this time, the hygrophiles were restricted to marshy parts, while euryhydric species inhabited the dry lake bottom.

A transition from terrestrial to marine tardigrade species can be observed when studying the fauna inhabiting maritime lichens (Morgan and Lampard, 1986; Kinchin, 1992*b*) (Figure 7.1). This is a particularly harsh environment, since the saxicolous lichens are subjected to insolation and the consequent extreme temperature fluctuations; they receive very little fresh water and are frequently subjected to salt spray.

Marine habitats

Marine tardigrades can be divided into three major ecological groupings (Renaud-Mornant, 1982*a*; Grimaldi de Zio *et al.*, 1983).

(i) Species restricted to the organic slime occurring on algae (e.g. *Styraconyx sargassi* on the invasive seaweed *Sargassum* spp.), on the plates of barnacles (e.g. *Echiniscoides sigismundi*) and other substrata, including those which are ectoparasites on marine invertebrates (e.g. *Tetrakentron synaptae* on the holothurian *Leptosynapta galliennei*). This species displays a number of morphological adaptations to ectoparasitism, including dorso-ventral flattening and development of extra hooks on the claws, and there is evidence that the cells of the holothurian are punctured by the tardigrade which then sucks out the cell contents (Van der Land, 1975; Kristensen, 1980). Other species have been found in association with other invertebrates, but it is not clear whether these are casual relationships or frequent parasitic/commensal relationships. They include *Actinarctus doryphorus* on the echinoid *Echinocyamus pusillus* (Schulz, 1953); *Pleocola limnoriae* on the isopod *Limnoria lignorum* (Cantacuzène, 1951), and *Echiniscoides sigismundi* on the bivalve *Mytilus edulis* (Green, 1950).

(ii) Interstitial species which can be found in the top few centimetres of sand in the littoral zone. These mesopsammic species are represented mainly by members of the genus *Batillipes* (Figure 7.2). One or more species of *Batillipes* appear to be present in the intertidal zone of nearly all exposed sandy beaches that have been studied, with investigations having concentrated on both sides of the North Atlantic (e.g. Renaud-Debyser, 1959; McGinty and Higgins, 1968; Pollock, 1970*b*; Lindgren, 1971; Harris, 1972; Martinez, 1975). In general, these

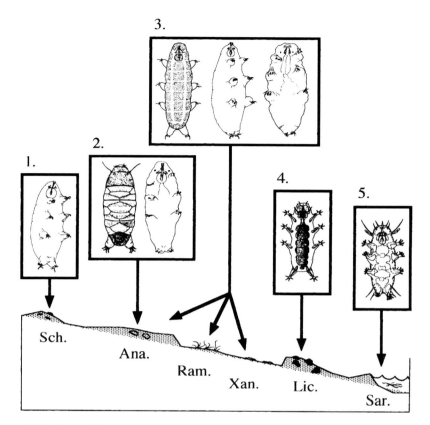

Figure 7.1. A possible zonation of tardigrades associated with littoral cryptogams
(1) *Hypsibius dujardini*; (2) *Bryodelphax parvulus* and *Macrobiotus echinogenitus*; (3) *Ramazzottius oberhaeuseri*, *Isohypsibius prosostomus* and *Milnesium tardigradum*; (4) *Styraconyx haploceros*; (5) *Styraconyx sargassi*. Abbreviations used: Sch., *Schistidium*; Ana., *Anaptychia*; Ram., *Ramalina*; Xan., *Xanthoria*; Lic., *Lichina*; and Sar., *Sargassum*. Based on Ramazzotti and Maucci (1983), Morgan and Lampard (1986) and Kinchin (1992b). Habitus drawings from Ramazzotti and Maucci (1983), Bertolani (1982b) and Dastych (1988).

studies have shown that seasonal populations of tardigrades (with peaks typically in spring and autumn) are most abundant in the top few centimetres of sand, particularly around the low tide level. While grain size is of importance to the intertidal meiofauna in general, by affecting the circulation of water and availability of O_2, some studies suggest that it is not a very important factor in determining the distribution of tardigrades. Where more than one species occurs, there does seem to be a zonation (Figure 7.3), possibly as a result of different feeding strategies, suggested by observed differences in buccal apparatus and the colour of gut contents (Pollock, 1970a). These animals probably graze upon various epipelic protoctists. Such macro-zonations are more easily observed in the psammolittoral environment as it is reasonably homogeneous, allowing environmental gradients to be identified. This is more difficult in the patchy environment occupied by bryophilous species.

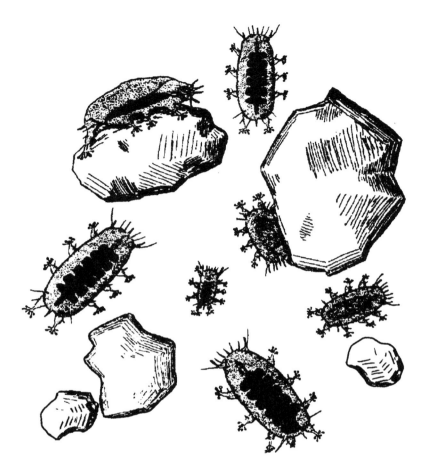

Figure 7.2. *Batillipes mirus* in its natural habitat
Redrawn from Marcus (1929b).

(iii) Deep sea benthic species which account for the majority of known marine species. Little is known about their biology and many species are represented by only a few specimens because of the difficulties in collecting these animals. This group includes those with a vermiform body outline which are thought to inhabit muds, e.g. the Coronarctidae (Figure 7.4). This shape, combined with short appendages, is thought to be advantageous to animals digging in mud and so aid limited active dispersal of the species. Also in this group are those with highly developed appendages and cuticular projections, which are more common in and around coarse sediments and sands (Figure 3.4). The increase in surface area created by these projections is thought to aid passive dispersal, as the animals are taken in water currents and possibly transported over large distances. It has also been hypothesized, by Higgins (1977b), that marine tardigrades may be dispersed as part of the 'fouling community' on the hulls of ships, which often includes the barnacles and *Enteromorpha* with which species such as *Echiniscoides* cf. *sigismundi* are frequently associated.

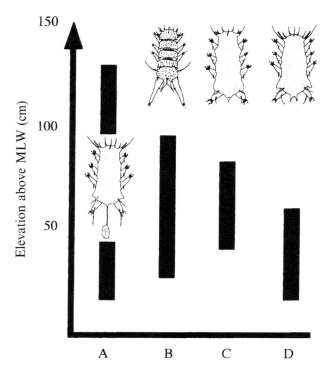

Figure 7.3. Observed zonation of psammolittoral tardigrades at Crane's Beach, Woods Hole, Mass., U.S.A.
(A) *Batillipes bullacaudatus*; (B) *Stygarctus granulatus*; (C) *Batillipes pennaki*; and (D) *Batillipes dicrocercus*. Based on Pollock (1970b).

Overall body shape of marine tardigrades, along with the detail of their claw structure is, therefore, an indicator of the animals' life-style and preferred habitat (Renaud-Mornant, 1982a; Grimaldi de Zio *et al.*, 1982b; 1984).

Habitat preference

There have been few investigations of the positions occupied by tardigrades within a moss cushion. When saturated, the water pockets in the leaf axils of a moss cushion form an interconnected spiral network, allowing the fauna to migrate vertically (Hallas, 1975a). Overgaard-Nielsen (1948) has described a vertical zonation within a moss cushion composed of three layers:

- a-layer

The productive green leaves at the top of the cushion. This is the zone provided with most oxygen and sunlight, but is also most prone to rapid desiccation.
- b-layer

The brown layer of degenerate leaves and stems, often described as the stem layer or litter layer.

- c-layer
The rhizoid layer and the soil trapped within it. This layer provides the most stable environment and is the last to dry out.

Hallas (1978) has constructed a model to explain the observed vertical distribution of some of the most commonly observed bryophilous eutardigrades. This is based on the cycle of degeneration within a moss cushion. As the cushion grows, the a-layer slowly dies back and forms the b-layer. After further decomposition, this becomes the c-layer. Therefore, the tardigrades found in the c-layer are true soil-dwelling species; those in the b-layer are litter species and those in the a-layers are the most specialized bryophilous species (Figure 7.5).

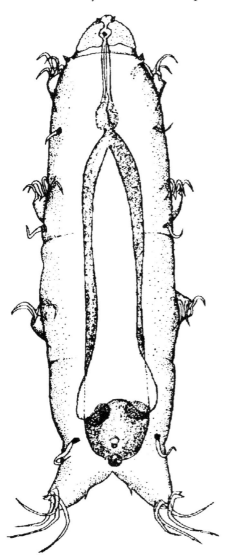

Figure 7.4. A vermiform deep sea tardigrade, *Coronarctus tenellus*
This animal has been recovered from the meiobenthos from the Indian and Atlantic oceans at depths between 400 and 3700 m. Marine species that have been recovered from more than one ocean and their marginal seas are regarded as cosmopolitan (Sterrer, 1973). Reduction in the size of the legs and cephalic appendages in this species are adaptations to life in an interstitial environment. Redrawn from Renaud-Mornant (1974).

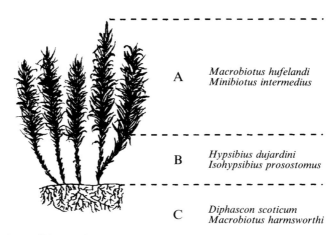

Figure 7.5 A possible vertical zonation of bryophilous tardigrades within a moss cushion
Reproduced from Kinchin (1989b), modified after Hallas (1978). See text for details.

However, it is not clear whether these species maintain their vertical position whatever the environmental conditions, or whether they undergo vertical migrations similar to those observed for some of the more desiccation-tolerant bryophilous nematodes (Overgaard-Nielsen, 1948). In those xeric moss cushions in which dew formation is the major source of water, elements of the bryofauna may undergo diurnal vertical migrations to maximize the periods of optimum humidity. Alternatively, the animals may maintain their positions and undergo daily anhydrobiosis. It is possible that both strategies are exhibited by different tardigrade species — even within the same moss cushion. Wright (1991) has shown that some tardigrades are more likely to migrate than others in a community in a cushion of *Grimmia pulvinata*. While *Macrobiotus richtersi* apparently migrates towards the c-layer along with the receding water channels, *Echiniscus testudo* remains in the a-layer while it dries out. The degree of desiccation tolerance exhibited by a given species has been linked to its morphology — particularly the morphology of the cuticle (Wright, 1988a; 1989b).

The species in Table 7.1 have been divided up among the categories on the basis of the frequency with which they appear in the different environments. This is also the case for the data on altitude and substratum preference described below. However, the accuracy of such a classification is a reflection of the number of observations made and even where a large number of observations have been made, eventual discrepancies between entries in the lists and observations in nature are bound to occur.

Other environmental factors may influence habitat preference and mask factors, such as bedrock pH (Table 7.2). For example, *Diphascon scoticum* has been found to be a common component of urban tardigrade communities. Its occurrence has been linked to atmospheric polutants (particularly SO_2) which effectively reduce the pH of the substratum (Meininger et al., 1985).

Table 7.1. The preferred moisture regimes of common tardigrade species

Hydrophilous	Typical of moss group 1	e.g. *Dactylobiotus* spp.
		Pseudobiotus spp.
Hygrophilous	Typical of moss groups 2 and 3	e.g *Hypsibius dujardini*
		Minibiotus intermedius
Xerophilous	Typical of moss groups 4 and 5	e.g. *Echiniscus* spp.
		Ramazzottius cf. *oberhaeuseri*
		Milnesium tardigradum
Euryhydric	Typical of moss groups 1–5	e.g. *Macrobiotus* cf. *hufelandi*
		Macrobiotus cf. *harmsworthi*

Table 7.2. Acid/base bedrock: preferences of some of the more common limno-terrestrial species of tardigrade

Eucalciphil	Exclusively on carbonate/alkaline rocks	e.g.	*Echiniscus granulatus*
			Echiniscus testudo
			Isohypsibius elegans
			Richtersius coronifer
Polycalciphil	Usually oncarbonate/alkaline rocks	e.g.	*Ramazzottius* cf. *oberhaeuseri*
			Macrobiotus areolatus
Mesocalciphil	No observed preference for carbonate or non-carbonate rocks	e.g.	*Macrobiotus* cf. *harmsworthi*
			Macrobiotus cf. *hufelandi*
			Milnesium tardigradum
			Minibiotus intermedius
			Hypsibius dujardini
Oligocalciphil	Usually on non-carbonate/acidic rocks	e.g.	*Calohypsibius ornatus*
			Diphascon scoticum
Acalciphil	Exclusively on non-carbonate/acidic rocks	e.g.	*Echiniscus blumi*
			Hebesuncus conjungens
			Diphascon angustatum
			Macrobiotus spectabilis

Data taken form Dastych (1988)

Eucalciphil species may, therefore, be found to be rare in heavily polluted areas. The importance of substratum pH as an environmental factor is thought to diminish as life conditions become drier, e.g. category 5 mosses (Hofmann, 1987), and other factors then tend to exert a greater influence on tardigrade populations. Evidently, other (sometimes unidentified) factors are influencing the distribution of tardigrade species. For example, Meininger and Spatt (1988) observed that the oligocalciphil species, *Diphascon scoticum* was more commonly found near a road polluted by alkaline dust than further away from the road where the substratum pH was lower. It is not clear what other environmental factors were influencing this distribution pattern.

Habitat preference of marine species is much less well understood, though some free-living species are known to be associated with specialized microhabitats, e.g. the association of *Moebjergarctus manganis* with deep sea manganese nodules in the South Pacific (Bussau, 1992). It is clear that many species have a preference for a certain type of sediment (sand or mud) as mentioned above, but their activities within these sediments are largely unstudied. Laboratory experiments have attempted to determine the effects of certain environmental variables on the position of some psammolittoral marine species within their sediments (e.g. Pollock 1975*b*), but further investigations, using a wider range of environmental variables on a range of species, are necessary before any generalizations can be made.

Distribution and dispersal

As a result of their small size, tardigrades are only capable of very limited active dispersal. Ramazzotti and Maucci (1983) calculated a maximum rate of progress of 17.7 cm/h for *Macrobiotus hufelandi* under optimal conditions. This is certainly sufficient for the animals to control their position within a moss cushion, as described above, possibly via saturated ectohydric channels on the moss stems. However, on the macroscopic scale, passive dispersal has a much greater influence on tardigrade distribution.

Studies on the passive dispersal of small limno-terrestrial organisms have shown that animals, such as rotifers and nematodes (which are of a similar size to tardigrades and are often found in the same microhabitats, see below), can be transported from one body of water to another by wind, or attached to soil and debris on animals such as arthropods, molluscs, frogs or birds (Maguire, 1963; Carroll and Viglierchio, 1981; Ramazzotti and Maucci, 1983). It is often assumed that the ability of tardigrades to dehydrate allows them to be transported in the anhydrobiotic state along with dry and dusty debris from the substratum; Kristensen (1987) has also observed *Echiniscus* spp. in raindrops or 'air plankton' after 'Föhn storms' in Greenland. However, some experimental studies, specifically on the transportation of tardigrades by wind, have not supported this view. In laboratory experiments, Sudzuki (1972) failed to observe the passive movement of tardigrades in air currents although smaller protozoa were moved in this way. Wright (1987) has suggested that wet conditions may favour tardigrade dispersal, by showing that the animals may be thrown into the air when rain drops impact upon the substratum. This would presumably only be responsible for movements of tardigrades over fairly short distances.

When observing *Cornechiniscus lobatus*, Ramazzotti and Maucci (1983) noted a distribution pattern of isolated dense aggregations. Similar distribution patterns have since been commonly observed (e.g. Kinchin, 1989a). In an attempt to explain this, three hypotheses were put forward by Ramazzotti and Maucci (1983). (i) That these rare tardigrades, localized in tiny areas, have numerous special needs and thus avoid large areas. They can, therefore, be described as stenoplastic species. (ii) That these are rare species in the region studied that have arrived there recently, carried by wind or animals. Local diffusion would be just beginning and would not have reached surrounding microhabitats. (iii) That the factors in (i) and (ii) are acting together — the animals are stenoplastic and recent arrivals.

Once a suitable microhabitat has been reached, its successful colonization by tardigrades is heavily influenced by the reproductive strategy of the species. For the genus *Ramazzottius*, the colonization by amphimictic and parthenogenetic strains has been studied in relation to the passive dispersal of tardigrades. Parthenogenetic strains have been found to be particularly successful in colonizing new territories starting from a single individual, for example, *Ramazzottius oberhaeuseri* (Bertolani et al., 1990). In contrast, the closely related species *Ramazzottius varieornatus* appears to have a very restricted distribution (Bertolani and Kinchin, 1993).

Numerous observations of tardigrades from a variety of heights above sea level have suggested that some species prefer certain altitudes, and in some instances may have a restricted range (e.g. Maucci, 1980; Ramazzotti and Maucci, 1983; Beasley, 1988). After an exhaustive study of the tardigrades of Poland, Dastych (1987a; 1988) has described five major groups of tardigrades, separated by their altitudinal distribution; these are summarized in Table 7.3.

The distribution of tardigrades at any altitude is primarily determined by abiotic factors associated with the microhabitat rather than altitude, moss species, exposure or geographical location, since the latter only affect the animals indirectly (Kathman and Cross, 1991). The precise environmental variables determining this distribution are not known, but probably include humidity and temperature. It is the effect on these factors by variation in altitude that will influence the tardigrade fauna. Any altitude-related pattern of tardigrade distrubution will, therefore, be distorted by the occurrence of sheltered microhabitats at high altitude and the nature of the bedrock.

On a smaller scale, it has been shown for corticolous species that the height of the moss above the ground influences the composition of its tardigrade fauna (Nelson, 1975). The factors determining this are not known, but a similar zonation of moss species and moss-inhabiting oribatid mites has also been described (Gjelstrup, 1979). Competition or predation could, therefore, be influencing the tardigrade fauna in addition to the moisture regime of the moss species, which tends to get drier with increasing height above the ground. In urban areas, the activities of dogs (urinating against tree trunks) has been shown to affect the moss species around the base of trees, forming the so-called 'canine

Table 7.3. Altitudinal distribution of tardigrades

Altitudinal group	Distribution centre (height above sea level)	Example species
Lowland species	(Below 200 m)	*Ramazzottius anomalus*
Upland species	(201–500 m)	*Echiniscus testudo*
Foreland species	(501–1000 m)	*Bryodelphax weglarskae*
Montane species	(Above 1000 m)	
Subalpine	In mountain forests	*Bryodelphax parvulus*
Mesoalpine	Below the tree line	*Richtersius coronifer*
Eualpine	Above the tree line	*Calohypsibius ornatus*
Tychoalpine species	All altitudinal zones	*Macrobiotus* cf. *hufelandi*
		Macrobiotus cf. *harmsworthi*
		Ramazzottius cf. *oberhaeuseri*
		Diphascon scoticum
		Milnesium tardigradum

Data taken from Dastych (1987a; 1988).

zone' of lush cryptogamic growth (Gilbert, 1989). It is not known how this affects the bryofauna.

From all these data, it is clear that some species are extremely euryplastic in their environmental requirements and seem to be found wherever there is sufficient water to permit activity for even a short time. For example, *Macrobiotus hufelandi* is found at all latitudes and altitudes, on any type of bedrock and in a wide variety of moisture regimes. Consequently, this species has been recovered from a wide variety of microhabitats all over the world and is often described as a 'cosmopolitan' species (however, there must be some doubt about the identification of specimens in the *hufelandi*-group, see Chapter 9). McInnes (1994) describes 22 species of limno-terrestrial tardigrades as cosmopolitan. In a survey of tardigrades in and around Nice (southern France), Séméria (1982) identified three species as 'holocosmopolitan' (*Macrobiotus hufelandi, Milnesium tardigradum* and *Ramazzottius oberhaeuseri*). These are cosmopolitan species which are also common in urban and synanthropic habitats. Other cosmopolitan species (*Macrobiotus harmsworthi* and *Minibiotus intermedius*) were not found in urban habitats, although they were observed outside the city, and they may be described as 'hemicosmopolitan'.

Population dynamics and densities

Populations of tardigrades from a variety of habitats have been studied. Observers have made quantitative estimates and, by periodic sampling, the dynamics of some populations have been elucidated. Some of these studies are summarized in Table 7.4.

Table 7.4. Population studies on tardigrades

Habitat	Species observed	Population peaks (troughs in square brackets)	Location	Authors
Moss	Hypsibius convergens	[Summer]	Germany	Marcus (1929b)
Moss	Macrobiotus hufelandi Pseudechiniscus pseudoconifer	Spring	Italy	Franceschi et al. (1962-63)
Moss	Macrobiotus hufelandi Echiniscus testudo	Spring	Wales	Morgan (1977)
Moss	Macrobiotus areolatus	N.D.	Italy	Ramazzotti (1977)
Moss	Echiniscus sp. Macrobiotus furciger Hypsibius dujardini Diphascon spp.	Early Summer	Antarctica	Jennings (1979)
Moss	Macrobiotus hufelandi	Spring and Autumn	England	Kinchin (1985)
Lichen	Milnesium tardigradum	N.D.	Austria	Schuetz (1987)
Soil and litter	Macrobiotus hufelandi Macrobiotus harmsworthi Diphascon scoticum	[Nov–Dec] and [May–Jun]	Denmark	Hallas and Yeates (1972)

Table 7.4. contd. Population studies on tardigrades

Habitat	Species observed	Population peaks (troughs in square brackets)	Location	Authors
Freshwater	*Dactylobiotus grandipes*	Autumn	U.S.A.	Schuster et al. (1977)
Freshwater	*Hypsibius* spp.	Dec–Mar	Antarctica	Everitt (1981)
Freshwater	*Isohypsibius saltursus*	Autumn	U.S.A.	Wainberg and Hummon (1981)
Freshwater	*Macrobiotus pullari* *Pseudobiotus augusti* *Hypsibius dujardini* *Isohypsibius* sp.	Spring	U.S.A.	Nelson et al. (1987)
Freshwater	*Pseudobiotus augusti*	Variable	U.S.A.	Kathman and Nelson (1987)
Marine sand	*Batillipes pennaki*	Spring and Autumn	Italy	De Zio and Grimaldi (1966)
Marine sand	*Batillipes* spp.	Spring/summer and Autumn	U.S.A.	Pollock (1970b)
Barnacle plates	*Echiniscoides sigismundi*	[Summer]	U.K.	Crisp and Hobart (1954)

N.D., no data provided.

Population densities vary enormously in both marine and limno-terrestrial populations. Mesopsammic species are usually found in low densities, but Thiermann (in Giere, 1993) reported a patch of sand near the low-water line on a Portuguese beach that yielded up to 3500 specimens of *Batillipes pennaki* per 100 cm^3. Moss cushions usually support a few tardigrades, but Morgan (1977) recorded densities of up to 823 per g (228 per cm^2) in an exceptionally rich moss sample from Swansea. Densities are not constant and temporal changes in populations have been correlated with a variety of environmental conditions:

- Food availability has been shown to affect some tardigrade populations. Hallas and Yeates (1972) recorded a positive correlation between numbers of *Macrobiotus harmsworthi* and its nematode prey. The midsummer decline of *Hypsibius convergens* observed in two populations by Marcus (1929b) was attributed to a depletion of the algal food source. The absence of tardigrades from certain freshwater beaches in North America may also be explained on the basis of negligible quantities of edible algae upon which the tardigrades may graze (Pennak, 1951).
- Franceschi *et al.*, (1962–1963) and Morgan (1977) demonstrated correlations between tardigrade numbers and climatic factors, particularly temperature and humidity. While both populations displayed peaks in spring, correlations with temperature and rainfall were not in agreement. The population in Swansea were negatively correlated with these factors 1—20 days before sampling, while the Italian population displayed a significant positive correlation. Morgan (1977) considers the observed differences to be a reflection of sampling durations and regional climatic differences. Certainly the climatic factors affecting temperate bryophilous tardigrade populations are poorly understood. Populations in xeric microhabitats, which may be dry for extended periods, may exhibit cycles with a periodicity greater than one year (Schuetz, 1987).

Bryophilous tardigrades are more common in polar and temperate regions than in the tropics (Nelson, 1991). Since the early systematic observations of the Scottish microscopist James Murray (1906; 1910), tardigrades have been observed to be a common component of the microfauna of Antarctica. Numerous later studies have confirmed this (e.g. Jennings, 1976; Everitt, 1981; Davis, 1981; Dastych, 1984; 1989; McInnes and Ellis-Evans, 1987; Usher and Dastych, 1987; Miller *et al.*, 1988; Ottesen and Meier, 1990; Utsugi and Ohyama, 1989; 1991; 1993; Schwarz *et al.*, 1993). Population studies in this region have shown that peaks occur immediately after ice has melted in the austral summer (Jennings, 1976; Everitt, 1981) (Figure 7.6). There is evidence of metabolic cold adaptation in polar tardigrades. The holoarctic species *Amphibolus smreczinskii* exhibits minute changes in metabolic rate within the temperature range 2–6°C — temperatures which represent summer temperatures in the moss habitats of Sptisbergen (Klekowski and Opalinski, 1989). A

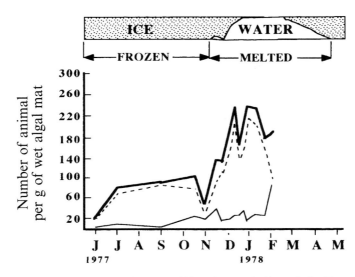

Figure 7.6. Graph of population changes of *Hypsibius* sp. in Deep Lake Tarn, Vestfold Hills, Antarctica during 1977 and 1978
Data expressed as number of animals per g of wet algal mat. Top bar indicates the depth of water frozen during the year. Thick solid line represents total tardigrades; dotted line represents small (juvenile) tardigrades; thin solid line represents large (adult) tardigrades. Redrawn from Everitt (1981).

similar picture emerges for temperate tardigrade species within a higher temperature range, suggesting that the animals adapt to local temperature conditions by developing a 'metabolic relative temperature independence' (Klekowski and Opalinski, 1989).

Other factors undoubtedly exert an influence on tardigrade populations, but have not been studied systematically. There is a suggestion that the length of the photoperiod may influence species with eyespots (Morgan, 1977; Kinchin, 1985). Population cycles in competitors, predators or parasites may also influence tardigrade populations.

Associated microbes

In addition to possible predation by some protozoans, as observed on nematodes (Doncaster and Hooper, 1961), tardigrades have regularly been observed to be infected by a particular protozoan, probably the symphoriont *Pyxidium tardigradum* (e.g. Van der Land, 1964; Morgan, 1977; Dastych, 1984). Infected tardigrades may carry one or a number of protozoa (Figure 7.7). These epizoites, feeding on bacteria, are not thought to be directly harmful to the tardigrades, although heavy infestations must present a considerable physical burden and hinder the movements of the tardigrade.

Fungi are very common in most of the microhabitats frequented by limnoterrestrial tardigrades. This can be seen if a hydrated moss cushion (or even just the washings from it) is left covered under laboratory conditions for a few days.

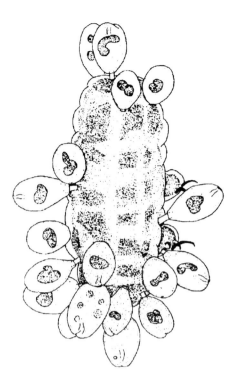

Figure 7.7. *Pyxidium tardigradum* (Protozoa) infecting *Ramazzottius* cf. *oberhaeuseri*
Redrawn from Van der Land (1964).

An extensive fungal mycelium will be found covering the material. This can be observed to hinder the movements of the tardigrades as their claws become intractably caught amongst the hyphae. Direct fungal attack of tardigrades also occurs. Fungal infections of tardigrades have been widely reported (e.g. Marcus, 1929*b*; Couch, 1945; Dreschler, 1951; Baumann, 1961; Richardson, 1970; Pohlad and Bernard, 1978; Dewel *et al.*, 1985; Dewel and Dewel, 1987; Barron, 1989; Kinchin, 1990*c*; Saikawa and Oyama, 1992; Saikawa *et al.*, 1991) and may be the most important parasites of tardigrades. Predatory fungi have also been observed to attack tardigrades and rotifers in moss and leaf litter (e.g. Barron *et al.*, 1990).

In general, fungal attack of bryophilous tardigrades is thought to reach a peak during the wettest months of the year (Morgan, 1977). Therefore, the dry periods may be important in reducing attacks by fungal parasites and explain why permanently moist mosses are often observed to have a poorer tardigrade fauna than adjacent dry mosses (e.g. Kinchin, 1992*b*). Under the optical microscope, fungal infections are often only obvious in their later stages when the parasite has invaded or replaced many of the host's internal tissues.

Members of the aquatic bryofauna (including nematodes, rotifers and tardigrades) are known to be infected by fungi of the genus *Harposporium*. [See Esser and El-Gholl (1992) for a review of the literature on this genus and a guide to the species within it.] *Harposporium anguillulae*, a common parasite of nematodes, has been observed by Saikawa *et al.* (1991) in *Macrobiotus* sp. (Figure 7.8). It is possible that infection is a result of the ingestion of conidia

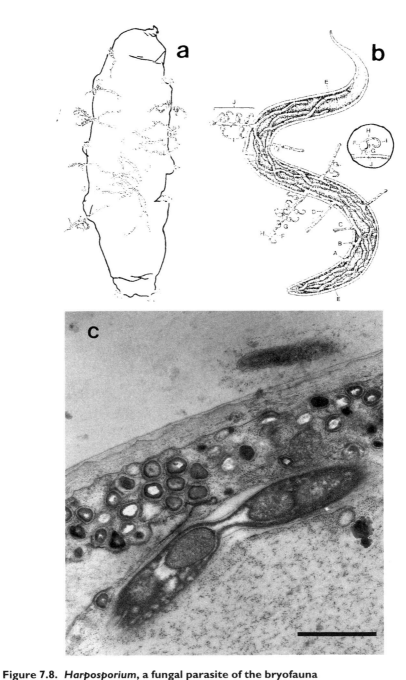

Figure 7.8. *Harposporium*, a fungal parasite of the bryofauna
(a) An infected tardigrade. Reproduced from Saikawa *et al.* (1991), with permission. (b) An infected nematode. Reproduced from Esser and El-Gholl (1992), with permission. (A) appressorium; (B) papillate bud; (C) conidiophore growth; (D) lateral buds; (E) mycelia; (F) phialide; (G) pedicel; (H) sterigma; (I) conidia; and (J) mature conidiophore. (c) TEM of a subcuticular fungal parasite, possibly inflated hyphal cells of *Harposporium* sp. Reproduced from Kinchin (1990b), with permission. Bar ≈ 1 μm.

rather than penetration of the cuticle, though both mechanisms are known from infected nematodes.

Recent electron microscopic examination of tardigrade infection by *Ballocephala sphaerospora* has shown how the fungus infects a new host by penetration of the tardigrade cuticle (Saikawa and Oyama, 1992). The conidia attach to the cuticle by an adhesive appressorium. After this, an infection peg penetrates the cuticle and produces an infection bulb in the body cavity of the tardigrade. The empty conidia remain on the surface of the animal.

Populations of the cosmopolitan tardigrade, *Milnesium tardigradum*, have been shown to be susceptible to parasitism by the fungus *Sorochytrium milnesiophthora* (Dewel *et al.*, 1985; Dewel and Dewel, 1987; 1990). Invading the tardigrade through the cuticle, the parasite first grows in the epidermis and later invades the body cavity. After consuming all the cellular contents of the tardigrade, the fungus produces zoospores to initiate infection of a new host. *Milnesium* eggs were observed by these authors, not to be vulnerable to fungal attack. Other inclusions have been observed in a number of eutardigrade species, particularly *Dactylobiotus dispar* and *Diphascon* spp. Elliptical structures with a barrel-shaped mid-section filled with cytoplasm and apparently empty hemisperes at each end have been termed 'X-bodies' by Van Rompu and De Smet (1991). These bodies measure 20 µm by 10 µm and may be of fungal or protozoal origin.

Bacteria and tardigrades are often found in the same microhabitat. Many tardigrades feed on bacteria. In addition, bacteria are often observed on the surface of tardigrades (Figure 7.9), sometimes in great densities. It is possible that some of these are utilizing materials secreted through the cuticular pores of some species.

Symbiotic bacteria have been observed in Florarctid marine tardigrades (Kristensen, 1984). The bacteria are restricted to the cephalic vesicles and their populations seem to be actively controlled by the tardigrades by occasional eversion of the vesicles to remove the bacteria. Secretions from the bacteria seem to pass across the tardigrade cuticle and may constitute a secondary energy source when particulate food is not available.

Bacteria have also been observed in the midguts of a number of eutardigrade genera (Figure 7.10), particularly *Ramazzottius*, *Hypsibius*, *Diphascon* and *Isohypsibius* (S.J. McInnes, personal communication). It is generally assumed that such intestinal floras are food items, although in some cases there is a suggestion that they may be symbiotic (Kinchin, 1989a; 1993). Virus particles have also been observed in tardigrade tissues, but their effects are unknown.

Associated microfauna and trophic relationships

As they are so rarely observed in their natural habitats, the feeding relationships of marine species (except those known to be parasitic) are understood less

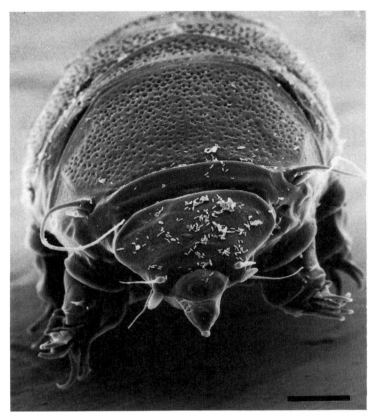

Figure 7.9. Anterior view of *Echiniscus horningi* showing epicuticular bacteria over the cephalic region
SEM courtesy of Diane Nelson. Bar ≈ 20 μm.

well than those of the limno-terrestrial species. Therefore, observation of gut contents is the most reliable indicator of their food preference. However, since the majority of a tardigrade's diet is composed of fluids, the colour of the gut contents is often all that can be used. The gut contents of *Echiniscoides sigismundi* vary in colour from red to green, suggesting a variety of food items, although fluorescence microscopy has indicated Cyanophytes to be a major component (Kronberg, 1983). Food coloration is also thought to change during digestion.

Other than a few non-specific phoretic relationships between tardigrades and bryophilous arthropods (Ramazzotti and Maucci, 1983), very few associations have been recorded for limno-terrestrial tardigrade species. Ectoparasitism is not often practical because of the tardigrade's requirement for a water film for continued activity. Endoparasitism has been recorded for one species from Puerto Rico. The heterotardigrade, *Echiniscus molluscorum*, is considered to be a parasite of the land snail, *Bulimulus exilis* (Fox and Garcia-Moll, 1962). The tardigrade was repeatedly recovered from the faeces of the

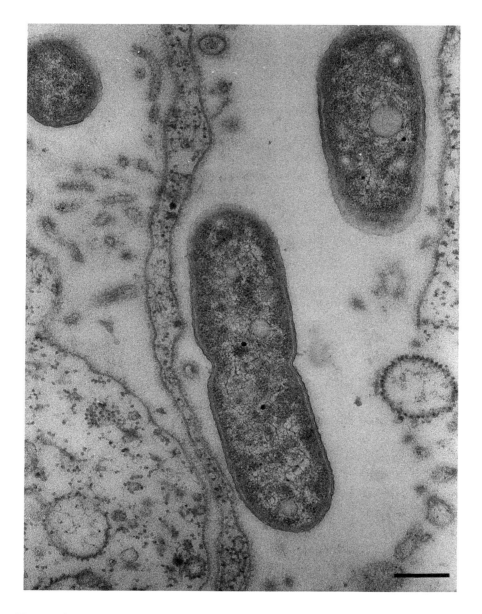

Figure 7.10. TEM of bacterial cells in a eutardigrade midgut
Bar = 0.3 μm.

snail, although it was absent from surrounding microenvironments, such as moss cushions. The exact nature of the association is not understood and it is possible that it has no nutritional basis.

In terms of studying trophic interactions, the bryophilous tardigrades pose less of a problem than the marine species and, consequently, more is known about them. The feeding preferences of many eutardigrades is correlated with the morphology of the buccal apparatus (Hallas and Yeates, 1972). Predators such as *Macrobiotus hufelandi*, *Macrobiotus harmsworthi* and *Milnesium tardigradum* are characterized by having a mouth opening that points directly forward; a short, wide buccal tube; robust stylets; and a large pharyngeal bulb. Algal feeders such as *Hypsibius dujardini*, have a more ventrally situated mouth and tend to have narrower buccal tubes and less-robust stylets. The observations of Hallas and Yeates on *Diphascon alpinum* suggest that the long, flexible buccal tube, characteristic of this genus, is an adaptation for grazing bacterial films by allowing the mouth opening to be positioned parallel to the surface of the substratum. Bryophilous tardigrades display a number of feeding strategies, but there is very limited evidence of bryophagy in tardigrades (e.g. Marcus, 1929*b*). Davis (1981) has calculated that no more than 0.04% of bryophyte production is consumed by the bryofauna of Antarctic tundra mosses (the proportion attributable to tardigrades is not clear). The epiphytic algae and bacteria growing on the surface of the moss leaves are probably a more important food source and are more likely to account for the chlorophyllous matter that is often observed in the midgut of bryophilous tardigrades.

Carnivory in tardigrades is also quite common. There are several published observations of nematodes being eaten by tardigrades (e.g. Hutchinson and Streu, 1960; Doncaster and Hooper, 1961; Hallas and Yeates, 1972; Iharos, 1975; Esser, 1990), and it has been suggested that the distrubution of some tardigrade species, such as *Macrobiotus harmsworthi*, is linked to the distribution of its nematode prey (Hallas and Yeates, 1972).

In many of the limno-terrestrial microhabitats frequented by tardigrades, a typical community of metazoan groups will be found in attendance, particularly in the drier microhabitats where the bryofauna are more specialized (Corbet and Lan, 1974). The relationships between the tardigrades and the other groups are known from only a handful of often anecdotal reports. It is clear, however, that a number of trophic relationships exist between the bryofaunal community members (Figure 7.11) and some knowledge of the associated microfauna is, therefore, of importance in the study of tardigrades. Some tardigrade species have also been observed to prey upon other tardigrades. *Milnesium tardigradum* is known to prey upon a variety of tardigrade species (Ramazzotti and Maucci, 1983) and this may explain its occurrence in low numbers in many tardigrade communities. *Macrobiotus richtersi* has been seen to prey upon *Hypsibius dujardini* and *Echiniscus testudo* (Wright, 1991).

Rotifers belonging to the Bdelloidea (Figure 7.12A) are often found with tardigrades in soil and moss cushions. The Bdelloidea are vortex feeders, creating currents in the surrounding water by the beating action of the cilia on their 'wheel organs'. This activity helps to locate active rotifers under the microscope. Like tardigrades, rotifers are only active when adequate moisture is present and many are able to utilize anhydrobiosis during periods of drought

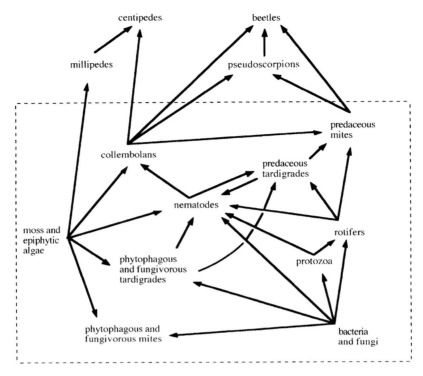

Figure 7.11. A food web of some of the trophic relationships within a bryosystem
Diurnal movements of those organisms within the box are restricted to the moss cushion. Arrows indicate flow of energy. Modified from Kinchin (1987).

(e.g. Gilbert, 1974; Ricci, 1987). Some species have also been observed to protect themselves from desiccation by constructing shells from soil debris and faecal material, or by taking refuge in the discarded shells of testacean amoebae (Donner, 1966; Wallwork, 1970). A number of tardigrade species are known to prey upon rotifers, including *Macrobiotus hufelandi* (Overgaard-Nielsen, 1948; Hallas and Yeates, 1972) and *Milnesium tardigradum* (Figure 13), and rotifer parts are commonly viewed in the midguts of predatory tardigrades. Christenberry and Mason (1979) report one tardigrade specimen in which the authors observe the remains of four rotifers and three other tardigrades. Predatory rotifers have also been observed with tardigrade remains within them.

In all microhabitats where tardigrades are found, nematodes (Figure 7.12D) are also typically found in large numbers. In the drier mosses, where the bryofauna is most specialized, the nematode fauna is dominated by members of the desiccation-tolerant genus, *Plectus*. Overgaard-Nielsen (1948) recognized three ecological groups of bryophilous nematodes, distinguished by the extent of their migrations within the moss cushion in response to desiccation.

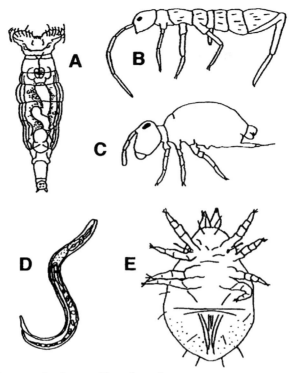

Figure 7.12. Commonly observed bryofaunal groups
(A) Bdelloid rotifer; (B) arthropleonid collembolan; (C) symphypleonid collembolan; (D) nematode; and (E) oribatid mite. Not to scale. Modified from Kinchin (1992b).

- Group 1

The largest group, whose members undertake pronounced vertical migrations from the c-layer to the a-layer (see Figure 7.5) when the moss is damp and return when the moss dries, e.g. *Plectus* spp.

- Group 2

Those whose migrations are less pronounced, e.g. *Aphelenchoides* spp.

- Group 3

Those which are restricted to the c-layer irrespective of the moisture conditions of the moss, e.g. *Dorylaimus* spp.

It will be found that Group 1 nematodes are early colonizers of moss cushions and similar fluctuating microenvironments, while Group 3 nematodes will be late colonizers. Anhydrobiotic nematodes are often found in mosses in a coiled condition (Demeure *et al.*, 1979; Kinchin, 1989b). This coiling is to slow the rate of dehydration, by reducing the exposed surface area of cuticle, and is analogous to tun formation in tardigrades. This is often accompanied by aggregation behaviour.

Arthropods are also represented in bryofaunal communities (e.g. Kinchin, 1990a; 1992b), but whereas tardigrades, nematodes and rotifers need to be

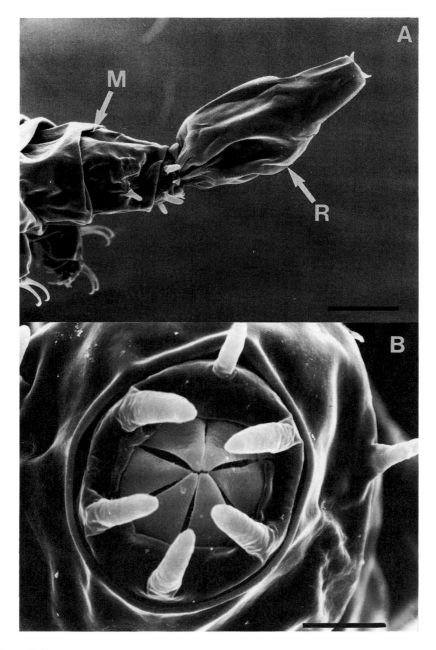

Figure 7.13 Lateral view of *Milnesium tardigradum* (M) ingesting a bdelloid rotifer (R) (top) with details of the buccal structure (below) showing the oral papillae
SEMs courtesy of Diane Nelson. Top bar ≈ 30 μm, lower bar ≈ 10 μm.

surrounded by a film of water if they are to move around, many bryophilous arthropods are air-breathing and so occupy the drier areas within a moss cushion. As a result of their different moisture requirements, these arthropods are most active at the times and in the places where the aquatic fauna tend to be inactive. At certain times in the transition periods between saturation and dehydration of the moss, both groups may be active and most trophic interactions will occur at these times.

In unfavourable drought conditions, adult arthropods cannot adopt the anhydrobiotic state used so successfully by the aquatic bryofauna, but the waterproof arthropodan cuticle means that many species are quite resistant to desiccation for short periods. Water loss is reduced still further by some animals by rolling up to reduce the exposed surface area (e.g. Phthiracarid mites). However, in extreme conditions, larger arthropods may be forced to leave the moss cushion in search of a more humid environment — an option not open to tardigrades, nematodes and rotifers which must await the return of suficient water. Among the arthropod groups often observed in association with mosses, the mites (Figure 7.12E) and collembolans (Figures 7.12B and 7.12C) are by far the most common.

Of the mites, members of the Oribatida are the most frequently observed in mosses and are often termed 'moss mites'. A number of species are thought to be bryophagous, feeding either on the contents of leaf cells or on bryophyte capsules (Gerson, 1969), while others probably feed on the epiphytic algae, fungi and bacteria on the surface of the moss leaves. Oribatids tend to be slow-moving animals, but a number of fast-moving, predatory Prostigmatid and Mesostigmatid mites may also be found in the bryosystem.

The other common arthropod group, the Collembola (spring tails) are primitive insects which lack a continuous waterproof cuticle, although their surface is strongly hydrophobic. The surface of the animal is covered in numerous water-repellent microtubercles (Lawrence and Massoud, 1973), between which is a permeable cuticle permitting gas exchange across the body wall. This prevents the animals from being submerged during periods of saturation, but makes them susceptible to dehydration and so humidity is a major factor determining their distribution within these microenvironments (Ashraf, 1971). This hydrophobic quality, combined with their ability to jump, makes live collembolans very awkward to handle (see Kinchin, 1990*a*). The unique and universal feature of collembolans is the possession of a ventral tube on the first body segment, posterior to the legs. Its function seems to be to absorb water and ions from the substratum into the body fluids. Collembolans have a wide range of feeding habits. Some species eat algae, fungi, bacteria or decaying plant remains, while others prey on juvenile nematodes.

Mosses not only supply food for the bryofauna, directly or indirectly, but also provide insulation against rapid changes in temperature and humidity, effectively extending the range of many arthropods into arid and subarctic regions (e.g. Hammer, 1966; Janetschek, 1967). Some arthropods may also participate in bryophyte spore dispersal and have also influenced bryophyte evolution by inducing the development of secondary plant substances, such as oxalic acid, and various alkaloids and tannins to discourage bryophagy. The effectiveness of these substances against tardigrade bryophagy has not been evaluated, although there is some evidence that echiniscids are able to eliminate bryophyte metabolites (Dewel *et al.*, 1992). Further examples of bryophyte/arthropod co-evolution are discussed by Gerson (1982).

8

Collecting and preserving

Collecting

The collection of specimens and extraction from their substratum is often one of the most time-consuming activities when studying tardigrades. The methods used will vary depending upon the substratum being sampled.

Various filtration and elutriation processes are commonly used to extract meiofauna (including tardigrades) from sediments. Animals can be extracted from sand by stirring the sand in a container of water and decanting the water through a sieve as soon as the sand settles (e.g. Schuster et al., 1977). For marine sands, Giere (1993) recommends subjecting the animals to a 'freshwater shock' to encourage them to release their grip on the sand particles. The classic Boisseau apparatus (Boisseau, 1957) described by Morgan and King (1976) is an efficient method for quantitative studies. This method has been variously modified by other authors (e.g. Uhlig et al., 1973); however, a simpler set up, using the same principle, has been described for the extraction of tardigrades from various substrata, including sediments and mosses (Kinchin, 1987) (Figure 8.1). The flow of water in this apparatus should be sufficient to agitate the moss, but not so great that 'large' particles are washed over into the sieve. The moss can be soaked in the inner column full of water to ensure thorough hydration of the animals before extraction. A little ethanol added to this water will help to loosen the animals from the plant material. Fine sieves (≈ 40 µm) will collect more material, including eggs, but will also collect more debris. Coarser sieves (≈ 90 µm) will collect less debris and so will be easier to sort through, but eggs will be lost.

Hallas (1975) has described a mechanical method for the extraction of tardigrades from soil or moss that involves mixing and centrifugation of the material. Greaves (1989) describes an extraction method which he has used to successfully obtain tardigrades fom soil, sand and leaf-litter. This is a modification of a method that is widely used to extract soil nematodes (Whitehead and Hemming, 1965). In essence, the sample is suspended over a

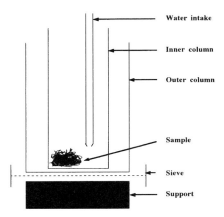

Figure 8.1. Apparatus for extracting tardigrades from their substratum
The entire apparatus stands in a sink with the water overflowing slowly. This should be left for at least an hour. Tardigrades can be recovered from the washings from the sieve under a binocular microscope. A narcotizing agent may be added to the water in the inner column. Redrawn from Kinchin (1987).

tray of water and held in tissue paper, through which the animals migrate over a period of 24 h. The water in the tray can then be filtered to recover the tardigrades.

Unless quantitative data are required, good yields of tardigrades can be obtained from moss by simply soaking the cushion in a petri dish of water and then squeezing the material in the water. Many tardigrades will be released into the water and can be recovered using a fine pipette.

Culturing

The maintenance of axenic cultures of tardigrades has proved to be more or less impossible. Those who have successfully cultured tardigrades have invariably maintained monoxenic cultures using either algae (Węglarska, 1957; Baumann, 1961; Blom, 1972) or nematodes (Sayre, 1969) as food items. These techniques are useful for observing tardigrade life cycles.

It is possible to maintain xenic cultures of an entire bryosystem, but unless extreme care is taken, the entire culture becomes over-run by fungal mycelia. To avoid this, the moss should be maintained at a low temperature ($\approx 5°C$) and the surrounding water should be changed daily. Blom (1972) has devised a method for the continuous changing of water, in a culture using absorbent wicks, which minimizes loss of specimens. However, the extraction of the tardigrades from the substratum still poses a problem with this sort of culture and it is questionable whether such complicated processes are worth the trouble of setting up and maintaining them.

Mounting

There are a variety of mounting media that can be used with tardigrades. One of the most convenient is Hoyer's medium. As it is an aqueous medium, fresh tardigrades can be mounted directly without having to first fix or dehydrate the specimens.

Hoyer's Medium contains: distilled water (50 cm^3); gum arabic (30 g); chloral hydrate[1] (150 g); and glycerol (20 cm^3). This is best made in a stoppered bottle. Add the gum arabic to the water and leave to stand for 24 h. Vigorous shaking or stirring should be avoided. The gum arabic will dissolve slowly. Light agitation with a glass rod may be necessary. Add the chloral hydrate to the mixture. This is very soluble, and the large volume will easily dissolve in the mixture. Leave to stand for another 24 h before adding the glycerol. It is possible for the gum arabic to form lumps in which case the mixture can be filtered through a sheet of silk into another bottle. Otherwise it can be used straight from the mixing vessel using a pipette or transferred into more conveniently sized dropping vials. The medium should be kept away from air or it will slowly become more viscous.

Slides will dry very slowly (the time varies with temperature) so they should be stored horizontally for several weeks to prevent movement of specimens and cover slips. Various stains, such as Lignin Pink, can be added to the mixture to enhance the specimens. However, chloral hydrate acts as a bleaching agent and so the natural colour of fresh specimens will be lost in a few hours. If this is to be recorded, specimens should be photographed immediately after mounting. Some shrinkage will be observed in eutardigrades immediately after mounting. However, the animals relax slowly and after about two weeks, shrinkage is reduced to 0–6% of the original fresh body length.

Before mounting, animals may be stored in a dehydrated state almost indefinitely, and there is anecdotal evidence, scattered throughout the literature, of animals that have been kept dry for many years and then successfully rehydrated. However, under some conditions, it seems that even dehydrated animals may have a limited 'shelf life' (Kinchin, 1993). There have also been attempts to devise protocols for the cryopreservation of tardigrades (e.g. Sayre and Hwang, 1975; Sugawara *et al.*, 1990), although this seems to have little or no advantage over dry storage.

---

[1]***Caution**: cloral hydrate is a toxic substance. Reagents should be clearly marked and kept out of the reach of children. Skin should be cleaned with copious amounts of soap and water after accidental contact. If ingested, seek urgent medical advice.*

9

Guide to common species

Introduction

While there are over 750 species of tardigrade now described, there are many fewer that will be regularly found by the interested observer. As most tardigradologists will not have access to the specialized equipment required to sample deep-sea environments, most of the species observed on a regular basis will be those from terrestrial, freshwater and littoral habitats. Scanning the literature will also show that there are a number of species which are sampled time and time again, while others have only been recorded once or twice (e.g. McInnes, 1994). This brief guide describes some of those species (or groups of species) which are the most likely to be observed by the interested microscopist. It is not possible to provide a comprehensive guide to species here; instead, enough information has been given to direct the reader to the appropriate sections of the scattered taxonomic literature as cited here and by others (e.g. Ramazzotti and Maucci, 1983; McInnes, 1994; Mackness, 1994).

Quantitative morphometric data

In the systematic study of eutardigrades at the specific level, many difficulties arise from the group's homogeneity; species often are differentiated by minute characteristics, which are difficult to observe and few in number. Another difficulty arises from the little use that can be made of quantitative characteristics, either because many structures are flexible and assume varied positions which make measurement impossible, or because other structures, such as the pharyngeal bulb, may be deformed by the pressure of the coverslip, or by the fixatives and mountants that are commonly used with tardigrades. Even when the distortion caused by a particular mountant is evaluated, it is apparent that mountants will vary from one batch to another. It is also apparent that workers use various mountants in their preparations (polyvinyl lactophenol and Hoyer's are seemingly the most popular). All this makes it very difficult to compare closely related species (particularly from published descriptions).

One of the most commonly quoted measurements in the description of tardigrades has been total body length (excluding the fourth pair of legs), which is of value in determining the minimum and maximum size in a particular species. However, in permanent preparations, the body of the tardigrade is often not fully relaxed and may exhibit a range of distortions, resulting in inconsistent measurements. I have elucidated the degree of distortion of body length, caused by my own batch of Hoyer's mountant, by measuring a number of specimens mounted in water with the coverslip supported by a human hair prior to mounting in Hoyer's. This revealed an immediate shrinkage in body length varying from 6% to 10%[1] of the fresh length (this will vary from one batch of mountant to another and also with the volume of water immediately surrounding the specimen when mounted, as this will dilute the mountant). However, specimens do relax in the mountant and, after two weeks, the degree of shrinkage was reduced to 0–6% of the fresh length. So, not only is it helpful to know which mountant was used in a description, but also how soon after mounting the animal was measured.

In an attempt to overcome the problems associated with the measurement of soft body parts, particularly in eutardigrades, a number of authors have used the length of the buccal tube (and sometimes its relationship to the body length) to separate species (e.g. Higgins, 1959; Morgan, 1976). Pilato (1981) has tried to standardize the measurement of eutardigrades and recommends measurement of the non-deformable sclerified structures (claws and buccal tube), since they are unaffected by mountants or coverslip pressure. These he relates to each other and to the overall body length in what he calls the percent ratio (Pt). Pt values are calculated as follows:

$$\text{Pt body length} = \frac{\text{Buccal tube length}}{\text{Total body length}} \times 100$$

$$\text{Pt stylet insertion} = \frac{\text{Distance to stylet support}}{\text{Total buccal tube length}} \times 100$$

$$\text{Pt claw length} = \frac{\text{Claw length}}{\text{Buccal tube length}} \times 100$$

Pilato acknowledges that detailed measurement of large numbers of claws can be a problem because of their small size, and because it can be very difficult to regulate their orientation in a permanent mount. Measurements in varying orientations would lead to inconsistent results. However, despite the difficulties, this technique has been used to separate species within the *Isohypsibius elegans* group (Pilato et al., 1982). Whether or not it might be useful to codify

---

[1] *Measurements were made upon specimens of* Ramazzotius varieornatus.

Pt ratios in the way that has been used for morphometric comparisons in the Nematoda (e.g. Nicholas, 1975) is a point for discussion.

A structure which is less problematic than the claws or total body length, is the buccal apparatus. Its greater size makes accurate measurement easier than that of the claws, and its rigidity makes it more reliable than total body length. There are two points of reference on the buccal tube which may be usefully compared: the total length of the buccal tube (excluding the bulb apophyses) and the point of the insertion of the stylet supports upon the buccal tube. When expressed as a Pt of the buccal tube length, the distance from the buccal cavity to the insertion of the stylet supports on the buccal tube shows comparatively little intraspecific variation, even with specimens of very different sizes, as may be encountered in a natural population (Figure 9.1). In an analysis of the *hufelandi* group, Bertolani and Rebecchi (1993) found that the Pt of stylet insertion exhibited the smallest degree of intraspecific variability and is, therefore, probably one of the most taxonomically useful measurements.

Coupled with the addition of quantitative data, the descriptions of the buccal apparatus of eutardigrades have become much more detailed over the years. Earlier descriptions have sometimes caused problems, and, as a result, may have led to authors describing different specimens of the same species as distinct taxa. An example of the consequences of inadequate descriptions is described by Dastych (1991) for *Hypsibius antarcticus* (see below). Only as the taxonomic importance of particular structures has become clear have descriptions (and their accompanying illustrations) become more exact.

With those eutardigrades exhibiting epidermal pigmentation, there is

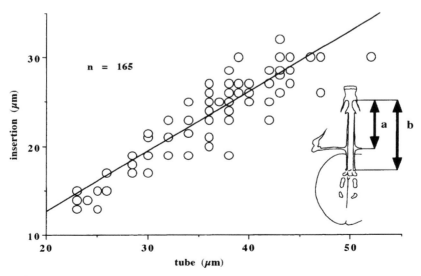

Figure 9.1. Plot of the Pt ratios for stylet insertion for 165 specimens from a population of *Ramazzottius varieornatus*, with (inset) detail of how the measurements were taken
(a) Stylet support; (b) total buccal tube length. Pt stylet insertion = $(^a/_b) \times 100$.

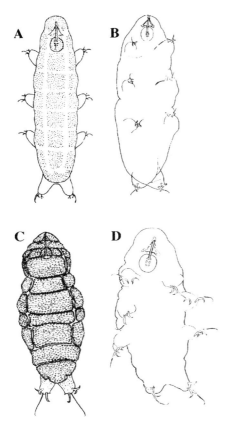

Figure 9.2. A comparison of illustrations of *Ramazzottius* spp. with and without pigmentation
(A and B) *Ramazzottius anomalus*. Reproduced from Ramazzotti and Maucci (1983) and Dastych (1979), with permission. (C and D) *Ramazzottius cataphractus*. Reproduced from Ramazzotti and Maucci (1983) and Dastych (1985), with permission.

apparent disagreement in the literature concerning whether or not to include the pigmentation in the illustrations of the animals (Figure 9.2). The pigmentation is a very obvious characteristic in the fresh animals, but the coloration soon disappears when the animals are mounted on slides. Consequently, when preparations are compared, the animals invariably appear unpigmented. Additionally, pigmentation seems to exhibit considerable intraspecific variation and so is quite unreliable as a diagnostic characteristic. Finally, by including pigmentation in illustrations, details of characteristics which are taxonomically more important (e.g. buccal apparatus) may be obscured. Therefore, while a description of the pigmentation of a specimen should be included in a description, either as a written explanation or as an additional drawing, an illustration of the mounted (unpigmented) specimen should be provided as it gives more information. This seems particularly sensible, since mounting is usually necessary to observe the fine detail required in species identification. Similarly, in the descriptions of echiniscids, details of cuticular texture have in some cases been omitted.

In accurate descriptions of species, it is also important to mention the absence of structures used in identification (e.g. microplacoids and cuticular bars). This has not always been recorded and has left room for ambiguity, since it has not always been clear whether the structures were absent or merely not recorded by earlier authors. Where such absences are recorded, e.g. Pilato (1971) stated that in *Isohypsibius deconincki* cuticular bars were absent on the claws of the first three pairs of legs, later observers can be sure that specimens exhibiting these characters are new varieties or species; e.g. *Isohypsibius hydrogogianus*, observed by Ito and Tagami (1993), is similar to *Isohypsibius deconincki* with the addition of cuticular bars on legs I, II and III. These cuticular bars are considered to be an important characteristic for species diagnosis (Pilato, 1975).

Macrobiotus cf. hufelandi

Macrobiotus hufelandi was the first species of tardigrade to be adequately described and has subsequently been recorded from a variety of habitats and environments worldwide (see Chapter 6). Described by Ramazzotti and Maucci (1983) as "the most common tardigrade, found everywhere", the animals are typically between 300 µm and 450 µm long (Figure 9.3). While juveniles are generally colourless, adults are opaque and may appear white, grey or brown. Eyespots are usually present. The animals have Y-shaped claws (Figure 9.4) with conspicuous accessory points and lunules. The basal piece of the claws are shaped like an inverted triangle, with the claws separated from the lunules by a thin peduncle (Bertolani and Pilato, 1988). The pharyngeal bulb has two macroplacoids, the first longer than the second and with median throttling. A microplacoid is also present. Eggs are deposited free, with ornamentation resembling inverted egg-cups or goblets (see Figure 5.3B).

The variation exhibited by specimens described as *Macrobiotus hufelandi* has long been recognized (e.g. Cuénot, 1932). Scanning electron microscope observations of *hufelandi*-type eggs by Grigarick *et al.* (1973) and Toftner *et al.* (1975) defined the structural variation of the egg ornamentation but, without parallel studies on the adult animals, species in the complex could not be separated. Further evidence that the *hufelandi*-complex consisted of a number of morphologically similar species was gathered by Bertolani (1973*a*; 1975; 1982*a*)

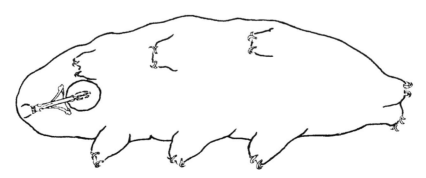

Figure 9.3. *Macrobiotus cf. hufelandi* **habitus**
Reproduced from Bertolani (1982*b*), with permission.

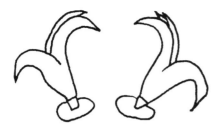

Figure 9.4. Detail of Y-shaped *hufelandi*-type claw

and Bertolani and Mambrini (1977) by karyological and morphological analysis of animals attributed to *Macrobiotus hufelandi*. Bertolani *(1982a)* concluded that animals previously described as *Macrobiotus hufelandi* consisted of more than one cytotype, but the 'new' species were not named.

Recent studies of eggs (Figure 9.5) and adults have led to a complete review of the status of the *hufelandi* group, including detailed descriptions of newly recognized species within the complex (Biserov, 1990a; 1990b; Bertolani and Rebecchi, 1993). The latter reference also includes a key to the adults of species within the *hufelandi* group and a second key to their eggs. It seems likely that as more *hufelandi*-type specimens are examined along the lines described by these authors, more species within this complex will be described and added to the 20 or so known to date. It is now clear that much of the variation described by earlier observers represents differences between the species within the group.

Macrobiotus hufelandi has been widely recorded from the British Isles (Morgan and King, 1976), but in the light of what has been said above, many of these records must now be in doubt and may not represent *Macrobiotus hufelandi (sensu stricto)*. There is a need for a thorough reappraisal of the *hufelandi*-group within the British Isles. When in doubt, it is always better to attribute a specimen to the *hufelandi*-group *(sensu lato)*, rather than incorrectly name it to the species level. One of the 'new' species within the group, *Macrobiotus sandrae*, has recently been recorded from southern England (Bertolani and Kinchin, 1993). As an additional complication, synonymy of *Macrobiotus hufelandi* with *Macrobiotus hibiscus* (De Barros, 1942) has not been ruled out.

Macrobiotus cf. *harmsworthi*

Specimens are typically 220–300 µm in length with a smooth cuticle (Figure 9.6). The oval pharyngeal bulb contains three macroplacoids and a large microplacoid. Claws are of the *hufelandi*-type with small lunules. *Macrobiotus harmsworthi* can be very difficult to distinguish from a number of other species within the genus, particularly in the absence of eggs. Similarities with specimens described as *Macrobiotus areolatus* and *Macobiotus richtersi* have caused particular confusion in the taxonomic literature. After detailed observations of the morphology of the egg shells of the three species, Hallas (1972) concluded that the characters which had been used to separate *Macrobiotus harmsworthi*

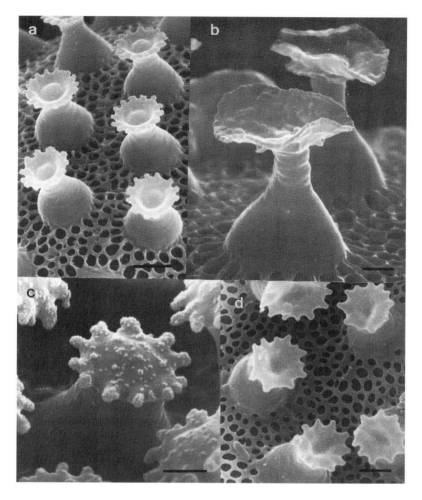

Figure 9.5. Details of the egg-shell ornamentation of species within the *hufelandi*-group
(a) *Macrobiotus hufelandi* (*sensu stricto*). Bar = 4 μm. (b) *Macrobiotus personatus*. Bar = 2 μm. (c) *Macrobiotus hyperboreus*. Bar = 2 μm. (d) *Macrobiotus sapiens*. Bar = 2 μm. SEMs courtesy of Vladimir Biserov.

areolatus and *Macrobiotus richtersi* were insignificant and that the three species should be assigned to *Macrobiotus harmsworthi* according to the rule of priority. This view has been rejected by others who consider all three species to be valid (e.g. Ramazzotti and Maucci, 1983; Pilato, 1975). Morgan, (1976) separated the three species as summarized in Table 9.1.

There are also similarities between *Macrobiotus cf. harmsworthi* and *Macrobiotus cf. hufelandi*. The two can be separated by the position of the stylet supports. In *Macrobiotus hufelandi* the attachment is close to or in contact with the pharyngeal bulb (never at a distance exceeding half the diameter of the buccal tube). In *Macrobiotus harmsworthi* the distance is equal to the

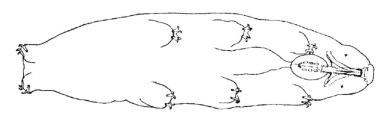

Figure 9.6. *Macrobiotus harmsworthi* habitus
Reproduced from Dastych (1988), with permission.

Table 9.1 Comparison of species within the *harmsworthi* group

Characteristic	*Macrobiotus areolatus*	*Macrobiotus harmsworthi*	*Macrobiotus richtersi*
Stylets	Small	Small	Large and robust
Pt of buccal tube diameter (external)	12.5–14.5%	10–14.5%	16.7–23.75%
Bulb apophyses	Small, closely associated with end of buccal tube	Large, closely associated with end of buccal tube	Large, distinct from end of buccal tube
Microplacoid	Generally absent (minute if present)	Small, greater than one-third the length of the third macroplacoid	Approx. one-half the length of the third macroplacoid
Eggs:	Figures 9.7A and 5.3A	Figure 9.7B	Figure 9.7C
Egg-shell projections	≤20 μm	≤25 μm	≤25 μm
Egg-shell areolations	Present	Absent	Present

diameter of the buccal tube. It seems doubtless that further species in the *harmsworthi* group will be described in the future. Separation of species will require detailed anatomical observations, analysis of egg morphology and Pt ratios of the buccopharyngeal apparatus. Pilato *et al.* (1991) have separated *Macrobiotus vanescens* from *Macrobiotus richtersi* using these criteria.

Milnesium tardigradum

Milnesium tardigradum (Doyère, 1840) (Figure 9.8) was one of the earliest species to be named and has subsequently been one of the most commonly observed and frequently studied eutardigrades. *Milnesium tardigradum* is a large, holocosmopolitan, carnivorous species with specimens occasionally

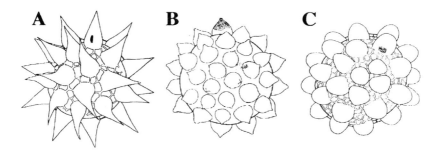

Figure 9.7. Eggs of *Macrobiotus areolatus* (A), *Macrobiotus harmsworthi* (B) and *Macrobiotus richtersi* (C)
Reproduced form Dastych (1988), with permission.

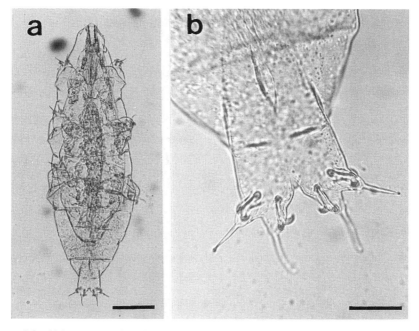

Figure 9.8. *Milnesium tardigradum*
(a) Ventral view of whole animal (bar = 100 μm); and (b) detail of claws on the fourth pair of legs (bar = 25 μm).

exceeding 1 mm in length, although 500 μm is more typical. The pharyngeal region of *Milnesium tardigradum* displays a number of interesting characteristics, including lateral and oral papillae and a pear-shaped pharyngeal bulb lacking placoids (see Figure 4.5). The mouth is surrounded by six triangular lamellae which close the aperture (see Figure 7.13B). The claws are unusual in that the primary and secondary branches are inserted upon the cuticle separately (Figure 9.8). The widest part of the body is typically around the position of the third pair of legs. Colour is very variable in this species, animals may appear colourless, brownish or pale pink (turning purple if treated with dilute potassium hydroxide solution). Eggs are smooth and spherical or oval. They may be colourless or brown with a diameter ranging from 70 µm to 110 µm. Typically, between six and nine eggs are laid in a caste exuvium, although as many as 18 may be observed in a single cuticle.

Milnesium tardigradum is particularly common in the drier terrestrial habitats, such as lichens, and has been recorded from the supralitoral lichen *Ramalina* spp. (e.g. Morgan and Lampard, 1986; Kinchin, 1990b; 1992b). The typical form of *Milnesium tardigradum* has a completely smooth cuticle; however, two other forms with cuticular variations are recognized by Ramazzotti and Maucci (1983): *Milnesium tardigradum* var. *granulatum* displays a granulation on the dorsal cuticle, particularly towards the caudal end of the body; *Milnesium tardigradum forma trispinosa* is characterized by the presence of three dorsal spines inserted near the caudal end of the body.

Both of these varieties were found intermingled with populations of the normal form. It seems probable that a species occurring as commonly as *Milnesium tardigradum*, and across such a wide geographical range, will exhibit a certain variation in its morphology. For 150 years after the description of the species, *Milnesium* was known as a monotypic genus. In recent years, additional species have been described: *Milnesium brachyungue* (Binda and Pilato, 1990), *Milnesium eurystomum* (Maucci, 1991) and *Milnesium tetramellatum* (Pilato and Binda, 1991). Care should, therefore, be taken when observing large eutardigrades with lateral and oral papillae, as other species within the genus may be awaiting discovery.

Ramazzottius cf. *oberhaeuseri*

Ramazzottius (formerly *Hypsibius*) *oberhaeuseri* has been widely recorded, particularly from drier microhabitats. The species has often been recognized on the basis of the distinctive colour banding on the dorsum, accompanied by the presence of *oberhaeuseri*-type claws (e.g. Mitchell, 1982; Greaves, 1991). However, coloration is highly variable and is not a good characteristic to use in species recognition (see above and Figure 9.2).

The *oberhaeuseri*-group was reorganized with the erection of the genus *Ramazzottius* by Binda and Pilato (1986). A number of species (formerly *Hypsibius*) were attributed to the new genus, and species subsequently described have been added. The claws of *Ramazzottius* are distinguished by the flexible tract separating the primary arm of the external claw from the base (Figure 9.9). Species attributed to this group include *R. anomalus*, *R. baumanni*, *R. cataphractus*, *R. edmondabouti*, *R. montivagus*, *R. semisculptus*, *R. subanomalus*, *R. theroni*, *R. thulini*, *R. tribulosus*, *R. valaamis* and *R. varieornatus*. Further species within this group are currently being described. The species are difficult to differentiate, and eggs are often required

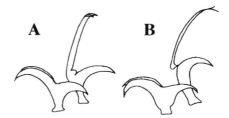

Figure 9.9. Detail of the *dujardini*-type (A) and *oberhaeuseri*-type (B) claws

to be assured of a species diagnosis, although they have not always been included in protologues (e.g. Seméria 1993). Pt ratios of the buccal apparatus are also of value in separating species (e.g. Bertolani and Kinchin, 1993) and should be included in future descriptions to aid comparison of specimens.

There is some confusion about the position of some species. For example, McInnes (1991) considers *novemcinctus* to be a member of the genus *Ramazzottius*, although Binda and Pilato (1986) originally considered that this species should remain in the genus *Hypsibius*. It is not clear whether the specimens described as *novemcinctus* by these authors are the same species or whether some of them are morphologically similar members of the genus *Ramazzottius*. Future records should indicate whether specimens are *Ramazzottius oberhaeuseri* (s. str.) or *Ramazzottius* cf. *oberhaeuseri* (s.l.), particularly if details of egg ornamentation or quantitative morphometric data are omitted.

Echiniscus testudo

Members of the genus *Echiniscus* are very common, particularly from the drier terrestrial habitats. *Echiniscus testudo* (considered to be the most common representative of the genus in the British Isles) has been widely recorded, especially in association with other species, and with members of the *hufelandi*-group (e.g. Morgan, 1977; Seméria, 1981; 1982; Mitchell, 1982; Bertolani and Kinchin, 1993). The animals are up to 360 μm long, typically with brown or red coloration. Red eyespots are present. Of the dorsal appendages, only spine Dd is present and is usually small (Figure 9.10). Eggs are laid in the exuvium. Two-clawed juveniles have only lateral filaments A and E. Filament C is developed in older animals. Two adult varieties of *Echiniscus testudo* are recognized: *Echiniscus testudo* var. *quadrifilis* with lateral filaments A, B, C and E; and *Echiniscus testudo* var. *trifilis* with lateral filaments A, C and E (i.e. lateral filament B is not acquired). Populations may be of one variety or the other, or may be mixed (Ramazzotti and Maucci, 1983).

Echiniscus granulatus

Echiniscus granulatus has been recorded widely within Europe, but only rarely in the British Isles. Usually associated with calcarious substrata, the animals are

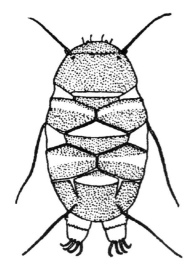

Figure 9.10. Dorsal view of *Echiniscus testudo* var. *trifilis*
Reproduced from Ramazzotti and Maucci (1983), with permission.

wide-bodied, generally 135–400 μm long with red coloration (see Figure 3.8). The dorsal cuticle has a very obvious texture of dots with rings around them (1.0–4.5 μm in diameter) similar to that described for *Echiniscus blumi*. Lateral appendages A, B, C and D are all filamentous (filament B is often absent and has only been observed once in British specimens). The terminal plate may be faceted in juveniles, but not in adults. Each of the fourth pair of legs has a dentate collar with 12–14 teeth. The internal claws have spurs, while the external claws are smooth.

Hypsibius dujardini

Hypsibius dujardini (Figure 9.11) has been found in numerous localities, usually from moss in damp or shady areas and from freshwater habitats. Animals are up to about 500 μm in length, usually with eyespots. The oval pharyngeal bulb contains two elongated macroplacoids, a microplacoid and well-developed apophyses. Claws are of the *dujardini*-type (Figure 9.9). *Hypsibius dujardini* is very similar to *Hypsibius convergens* in many respects. In *Hypsibius dujardini*, the macroplacoids are longer and the microplacoid is more obvious than in *Hypsibius convergens*.

Hypsibius cf. *antarcticus/arcticus*

The status of these polar *Hypsibius* spp. has recently been reviewed by Dastych (1991) as the original and subsequent descriptions of these species have caused considerable confusion in the literature and it seems probable they have often been misidentified. Descriptions of the two species in Ramazzotti and Maucci (1983) differ from each other only in minor details — both are described as having eyespots, smooth cuticle, two macroplacoids (with the first longer than the second) and large double claws (Figure 9.12). It is also clear that errors in

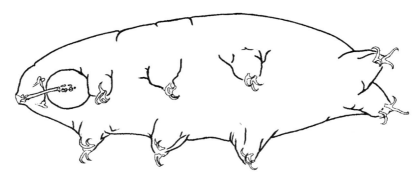

Figure 9.11. *Hypsibius dujardini* **habitus**
Reproduced from Bertolani (1982b), with permission.

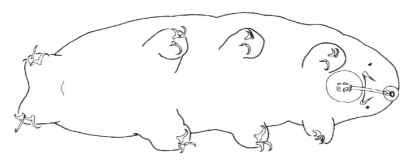

Figure 9.12. *Hypsibius antarcticus* **habitus**
Reproduced from Dastych (1991), with permission.

original illustration have been passed from one work to the next. This group also provides a good example of how tardigrade illustrations have changed over the past century (Figure 9.13). It is thought probable that a number of authors have mistakenly identified specimens of *Hypsibius arcticus/antarcticus*, for example as *Hypsibius dujardini* (Jennings, 1976).

After a detailed study of specimens, including type material where available, Dastych (1991) was unable to conclude whether *Hypsibius arcticus* and *Hypsibius antarcticus* were, in fact, specimens of the same species, with *Hypsibius arcticus* as a junior synonym of *Hypsibius antarcticus*, or whether the specimens represent two closely related species. Future detailed investigations similar to those described above for the *hufelandi*-group will be required in order to settle the problem. In the mean time, it is probably necessary to designate any possible specimens as *H*ypsibius cf. *antarcticus*.

Diphascon scoticum

Diphascon (Adropion) scoticum has been widely recorded. The animals are generally 215–370 μm in length with a slender body (Figure 9.14). Three long,

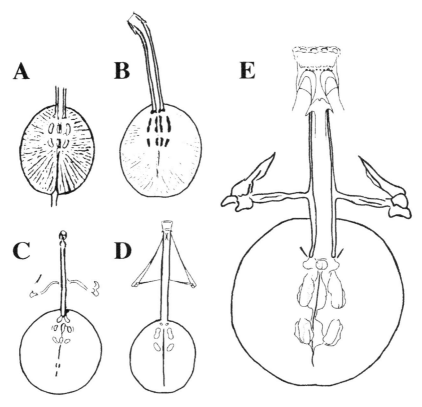

Figure 9.13. *Hypsibius antarcticus* **pharyngeal apparatus**
As drawn by: (A) Richters, 1908; (B) Petersen, 1951; (C) Robotti, 1972; (D) Maucci, 1986; and (E) Dastych, 1991. Reproduced from Dastych (1991), with permission.

thin macroplacoids are present, usually increasing in length from first to third. Sometimes the macroplacoids are of equal length. Microplacoid and septulum are usually present. The posterio-dorsal apodeme is absent. The average Pt of the stylet support insertion is approx. 74 (Pilato, 1987). The anterior apophyses for the insertion of stylet muscles are in the shape of semilunar hooks. Cuticular bars are present on the first three pairs of legs. The claws vary in size, but not shape, from leg to leg. There has been some problem with recognition of the simplex stages of *Diphascon scoticum* and its possible confusion with other species, or species within the genus *Itaquascon*.

A number of subspecies have been proposed. *Diphascon scoticum bicorne* is considered by Ramazzotti and Maucci (1983) to be a separate species, *Diphascon bicorne*, whereas *Diphascon scoticum ommatophorum* is considered to be a valid subspecies of *Diphascon scoticum* by the same authors. The subspecies *ommatophorum* is similar to the normal variety with the addition of eyespots.

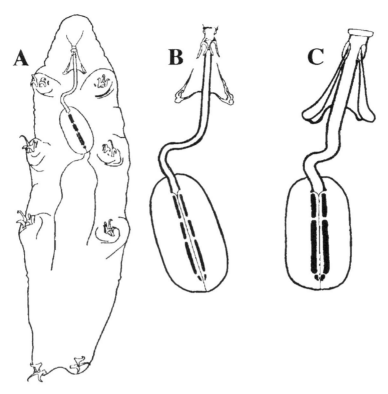

Figure 9.14. *Diphascon scoticum* habitus (**A**) with detail of the pharyngeal apparatus of *Diphascon scoticum* (**B**) and *Mesocrista spitsbergense* (**C**)
Reproduced from Dastych (1988), with permission.

Mesocrista spitsbergense

Specimens of *Mesocrista spitsbergense* (formerly *Diphascon*) are typically 240–260 µm in length, but can be longer. There are two macroplacoids (the second is two or three times longer than the first) and a microplacoid present (Figure 9.14). The average Pt of the stylet support insertion is approx. 87 (Pilato, 1987). The anterior apophyses for the insertion of stylet muscles are in the shape of wide, flat ridges. Eyespots are absent.

Richtersius coronifer

Richtersius coronifer (formerly *Adorybiotus* and *Macrobiotus*) are large animals, up to 1000 µm long (Figure 9.15). They are often yellow or orange, but sometimes colourless. Large eyespots are present. The cuticle is smooth with numerous pores. The pharyngeal bulb features two short, wide macroplacoids. Microplacoid is absent. The claws are of the *hufelandi*-type — with enormous dentate lunules, each with 10–18 teeth. The eggs are laid free and are ornamented with spikes; similar to those of *Macrobiotus islandicus*, but with larger

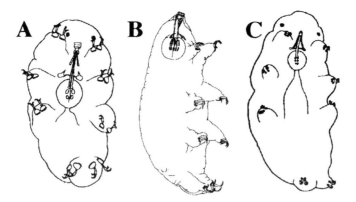

Figure 9.15. *Richtersius coronifer* **(A)**, *Dactylobiotus dispar* **(B)** and *Minibiotus intermedius* **(C)** habitus
Drawings reproduced from Bertolani (1982b) and Dastych (1988), with permission.

spikes (*Richtersius coronifer* ≤20 µm; *Macrobiotus islandicus* ≤11 µm). *Richtersius coronifer* was transferred to the genus *Richtersius* by Pilato and Binda (1987b; 1989) on the basis of differences in the buccopharyngeal apparatus.

Dactylobiotus dispar

Formerly *Macrobiotus*, *Dactylobiotus dispar* is an aquatic species of large size (typically up to 500 µm). The cuticle is usually smooth, but may be finely granulated. Eyespots are present. In some specimens, two dorso-lateral humps have been observed between the third and fourth pairs of legs. Two long, slender macroplacoids are visible in the pharyngeal bulb, the first twice as long as the second (Figure 9.15). Microplacoid is absent. The claws are very large and are connected by a chitinous bar (Figure 3.13C), which is particularly visible in the fourth pair of legs. Eggs (approximately 70 µm in diameter) are laid free and are ornamented with small, separated cones (4–5 µm in height). In the absence of eggs, the animals are difficult to separate from *Dactylobious ambiguus*.

Minibiotus intermedius

The genus *Minibiotus* was erected by Schuster *et al.* (1980) and, although rejected by Ramazzotti and Maucci (1983), who retained the species in the genus *Macrobiotus*, is now generally recognized as a valid genus (e.g. Bertolani, 1982b; Nelson 1991). *Minibiotus* (Figure 9.15) is distinguished from *Macrobiotus* by the short circum-oral papullae in the former and the elongated lamellae in the latter. *Minibiotus intermedius* is typically 125–250 µm in length. The body is white with a brown pigmentation. Large eyespots are present. A spherical pharyngeal bulb displays three roundish macroplacoids of approximately equal length and a tiny microplacoid. The claws are very small with large accessory spines on the primary branches and relatively large lunules. The eggs (38–45 μm in diameter) are ornamented with structures resembling hollow domes (3–3.5 μm in height).

Batillipes spp.

There are a number of species within the genus *Batillipes* which are commonly found in marine environments. The digits of this genus end in variously shaped discs, 'spoons' or 'shovels' (Figure 9.16). The length and arrangement of the digits is variable. The longest are dorsally situated, with the short and intermediate digits positioned laterally or ventrally so they can reach the surface of the sand grains upon which the animals walk. The general arrangement of the digits on the first three pair of legs in any given specimen shows little variation. However, the digits on the fourth pair of legs exhibit a different orientation as these limbs are only used in backward movements. In addition to the details of the digits, the caudal appendages are also an aid to species recognition (Figure 9.17). While this is a distinctive structure in many *Batillipes* species, McKirdy (1975) warns that the degree of intraspecific variation is not clear, and so other characteristics should also be taken into account, including the morphology of cephalic appendages, leg spines, buccal apparatus and the female gonoporal field. Various ratios have been employed in the morphological description of *Batillipes* spp. However, Villora-Moreno and de Zio Grimaldi (1993) consider the ratio of 'appendage size/body size' to be an unreliable aid to taxonomic descriptions as body growth and appendage growth are independent and may occur at various rates relative to each other. These animals are particularly common in psammolittoral environments.

Echiniscoides cf. *sigismundi*

Echiniscoides sigismundi has been widely recorded from littoral habitats, particularly associated with *Enteromorpha* or barnacle plates (Crisp and Hobart, 1954; Pollock, 1975c). These are robust animals, varying between 100 μm and 350 μm in length, with pronounced transverse folding of the dorsum and large simple claws. Eyespots are often observed. Obvious colouration of the animals results from their gut contents, varying from green to brown. Crisp and Hobart (1954) suggested that *Echiniscoides sigismundi* may compete with mites that are often found in the same microhabitats. However, densities of up to 16 animals

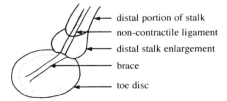

Figure 9.16. Detail of the *Batillipes*-type foot
Redrawn from McKirdy (1975).

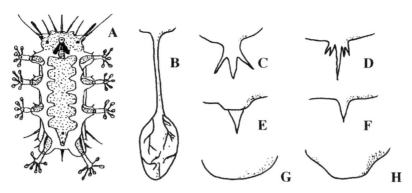

Figure 9.17. (A) *Batillipes mirus* (ventral view), with detail of the caudal processes of the British *Batillipes* spp, (B) *Batillipes bullacaudatus*; (C) *Batillipes littoralis*; (D) *Batillipes phreaticus*; (E) *Batillipes pennaki*; (F) *Batillipes mirus*; (G) *Batillipes acaudatus*; (H) *Batillipes tubernatis*
Redrawn from Pollock (1971).

per barnacle have been observed from Brighton on the south coast of England where mites, particularly *Isobactrus ungulatus* (Halacaridae), were also observed in great numbers (I.M. Kinchin, unpublished work).

Illustrations of *Echiniscoides sigismundi* have changed as emphasis upon various taxonomic characters has shifted over the years. Early illustrators, such as Marcus (Figure 9.18B) were more concerned with suggesting the habits of the animal rather than detail of the internal anatomy. This is in part due to the small number of species that were known — gross morphology was enough to distinguish the animal from others that had been described. The illustrations by Cuénot (1932) (Figure 9.18A) and Morgan and Lampard (1986*b*) (Figure 9.18C) also suggest differences in approach, one emphasizing the internal detail, while the other concentrates on the external morphology. As a number of additional species within the *Echiniscoides* genus were described in the

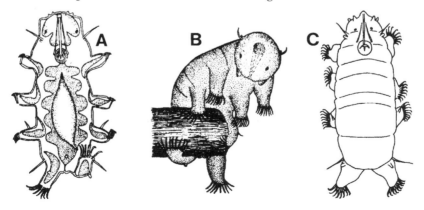

Figure 9.18. *Echiniscoides sigismundi* **habitus drawings**
Reproduced from (A) Cuénot (1932), (B) Marcus (1929*b*) and (C) Morgan and Lampard (1986*b*), with permission.

1980s, it became necessary to concentrate on details of the claws and buccal apparatus as shown in the precise illustrations by Kristensen and Hallas (1980) and Hallas and Kristensen (1982) (Figure 9.19). Future illustrations will need to show this degree of detail if they are to be of any value in describing species.

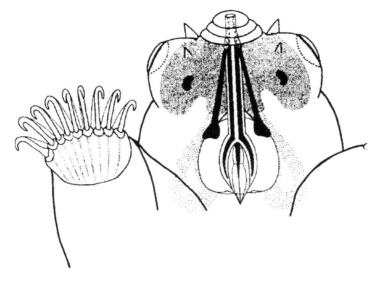

Figure 9.19. *Echiniscoides sigismundi* **showing detail of the buccal apparatus and claws**
Reproduced from Kristensen and Hallas (1980), with permission.

10

Future research

The scope for further research

There is ample scope for both the professional and the amateur to make valuable contributions to the body of knowledge concerning tardigrades. The designation of 'amateur tardigradologist' does not, in any way, infer inferior quality work — it should be remembered that the early microscopists and many of the great scientists of the past were amateurs by today's criteria (Brook, 1993). In today's economically fraught times, it may well be that the amateur has the advantage over the professional in that he/she is not pressurized by the 'publish or perish' or 'value-for-money science' philosophies. The amateur, unless a Rothschild or a Getty, may be confined by the cost of apparatus, but more important commodities are time and enthusiasm. While electron microscopy has revealed otherwise-hidden details of certain structures, the light microscope is still the most valuable item of equipment in the study of tardigrades.

Behaviour

Very little systematic investigation has been aimed at studying tardigrade behaviour. The few observations that have been noted by various authors are associated with mating (Chapter 4), cryptobiosis and tun formation (Chapter 5) and diurnal migrations (Chapter 6). Professor David Pye of London University has developed a method for studying phototactic behaviour in tardigrades (D. Pye, personal communication).

Animals are placed on a horizontal arena of wetted frosted glass which sits on a small carrier tray that slides smoothly over a greased sheet of plate glass. The arena can be manipulated to bring the tardigrade to the centre of the field of view of a binocular microscope, as defined by eye-piece cross hairs. As the tardigrade moves, the arena is moved in the opposite direction to keep the animal centred. A pantograph attached to the carrier slide records these movements with a ball-point pen on a sheet of paper. The arena carrier is also attached, by paired parallelogram levers, to a bar fixed to the bench (Figure 10.1). This allows free movement in two dimensions, but prevents rotation of the arena. By keeping the animal in a fixed position as it walks, the length of the recorded track is independent of the size of the optical field. In addition, the evenness of the stimulating light field is not critical, since the subject does not experience any spatial irregularities that may exist within the beam. Also, by observing the animals at fairly high magnification, detailed behaviour, such as klinotaxis, can be noted.

Stimuli consist of collimated light beams directed horizontally or at a grazing angle onto the arena and across the field of observation. Standard colour filters can be used to change the wavelengths of light present. In many of the earlier observations of phototaxis in nematodes, the effects of heat from the light source were neglected (Croll, 1970). Consequently, much of the published data is of limited value for comparison. In experiments on responses to light, heat has to be adequately excluded by inserting heat-absorbing glass filters or water baths between the light source and the arena. In the field, however, it should be remembered that heat and light occur together in sunlight. This should be taken into account when relating *in vitro* findings to life in a moss cushion.

Earlier observations on photoresponses in tardigrades have given conflicting results (Ramazzotti and Maucci, 1983). Some species have displayed a positive response, others display a negative response. Variations in the light intensity also seem to be important. Future observations should not only note the species of tardigrade used, but also the age of the specimens; Baumann (1961) observed that juveniles of *Hypsibius convergens* display a positive phototaxis immediately after hatching which disappears after the second moult.

It is not clear whether the presence of eyespots in some tardigrade species in any way influences phototactic behaviour. In nematodes, for example, most species for which light-sensitive responses are known lack localized pigment or light-receptive organelles. Their responses are thought to be mediated through a dermal light sense (DLS) (Steven, 1963; Croll, 1970). However, DLS in transparent tardigrades is unlikely to be able to produce a directional response, but may be more important in responses to changes in the photoperiod. Comparative studies of the behaviour and ecology of species with and without eyespots could help determine the occurrence of DLS in tardigrades.

There will evidently be some kind of balance between the hazards and benefits of exposure to daylight for tardigrades. In some species, the balance may be biased in favour of exposure, while in others behavioural responses may have evolved which tend to avoid light. The response to light may also be affected by other environmental cues such as temperature, humidity and oxygen tension. Light may act as a token stimulus, facilitating spatial orientation whereby a tardigrade can distinguish between 'up' and 'down' and avoid conditions which result in desiccation or asphyxia. It has been suggested that

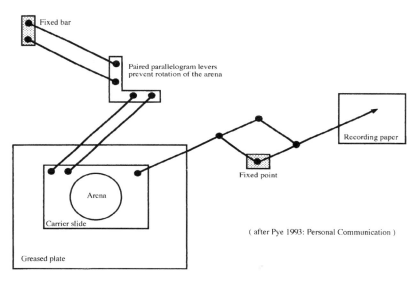

Figure 10.1. Apparatus used to investigate the locomotory behaviour of tardigrades (see text for details)

certain bryophilous species will undertake vertical migrations in a moss cushion to avoid repeated dehydration, while others are more likely to dehydrate than to modify their position within the moss cushion (e.g. Wright, 1991). This is based on the systematic observation of very few individuals of a small number of species. It is not known to which stimulus the animals are responding. It would be interesting to know whether the animals behave in the same way to changes in humidity under various lighting regimes, and thus determine the key factor.

Before detailed behavioural observations can be made on marine species, a method for their maintenance *in vitro* is required. This provides a significant barrier to the development of our understanding of these animals.

Biochemistry

There have been few histochemical or biochemical studies on tardigrades. These have tended to concentrate on the biochemistry of the cuticle sublayers (e.g. Greven and Peters, 1986; Volkmann and Greven, 1993) and enzyme activity in various internal organs (e.g. Raineri, 1982; 1985; 1987). These studies are still in their infancy and the histochemistry of other tissues needs to be examined to determine their physiology. The other aspect of tardigrade biology that has attracted the interest of biochemists is cryptobiosis (see Chapter 6). The role of trehalose as a membrane protectant seems to have been established, but the potential for using trehalose in the preservation of biological systems has still to be explored (see Roser and Colaço, 1993).

Ecology

Dispersal

An organism gains in fitness by dispersing its progeny as long as they are more likely to leave descendents than they would if they remained undispersed. The mode of dispersal used by limno-terrestrial tardigrades is not clear. Sudzuki (1972) supposed that dry disseminules were dispersed by wind, while Wright (1987) performed experiments to show that fresh animals could be dispersed in water droplets. Both methods, and the possible role of other animals in tardigrade dispersal, remain largely speculative. Further laboratory experiments, backed up by field observations, are required.

Specialized habitats

A number of insects, particularly weevils of the genus *Gymnopholus* from Papua New Guinea, have been observed to support a varied cryptogamic flora growing on their backs (Figure 10.2). This flora, comprising algae, fungi, liverworts, lichens, mosses and even fern gametophytes has in turn been observed to support a microfaunal community of collembolans, oribatid mites, nematodes and rotifers (Gressitt, 1966; 1969). Some species in this community are unique to this association. It seems likely that such a typical bryophilous community would also include a number of tardigrade species but, as yet, none have been recorded from this microhabitat.

Specialized moss communities, such as those growing on active travertine deposits (Pentecost, 1992) may also support unusual tardigrade faunas. Durante and Maucci (1972) found fossilized eggs of *Macrobiotus* cf. *hufelandi* in

Figure 10.2. *Gymnopholus lichenifer*, a weevil from the rain forests of Papua-New Guinea, supporting a collection of cryptogams on its dorsum
Drawing by Amy Bartlett Wright from Wilson (1992). Reprinted with permission.

travertines in Italy, suggesting that this is a suitable microhabitat for certain tardigrade communities.

Tardigrades have been recorded from phytotelmata, such as those in teasel leaf axils (Figure 10.3), although very few details are available (Maguire, 1959; Masters, 1967). Preliminary observations of the microfauna in teasel waters from southern England suggest that tardigrades are a rare component of this community (I.M. Kinchin, unpublished work). Observations from a wider geographical range would help to determine how common the tardigrade–teasel association is.

Urban tardigrades

There have been few systematic observations of urban tardigrades (e.g. Séméria, 1982; Meininger et al., 1985; Utsugi, 1986). There is a suggestion that both the heat-island effect and urban pollutants, such as sulphur dioxide, may affect the species distribution. Systematic monitoring of these environmental variables and a correlation of tardigrade species will establish any distribution patterns. If such relationships are shown, it may be possible to use a tardigrade-species-index to describe environmental quality. Other bryophilous meiofaunal groups have been shown to be sensitive to certain pollutants (e.g. Zullini and Peretti, 1986), and tardigrades are affected by certain known toxins (e.g. Barrett and Kimmel, 1972).

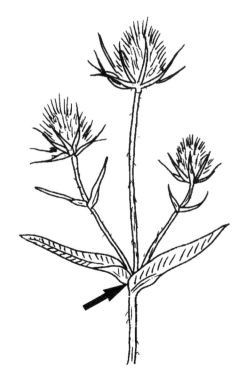

Figure 10.3. Flower head of the common teasel, *Dipsacus sylvestris*
Arrow indicates leaf axil.

Reproductive strategies

The reproductive strategies of very few tardigrade species (particularly marine species) have been determined. The variablity and significance of egg dormancy have not been established for tardigrades. Some species, such as *Ramazzottius* cf. *oberhaeuseri* have been observed to lay more than one type of egg (sculpted eggs, layed freely, and smooth eggs, layed in a cast exuvium) (Baumann, 1966). It is not clear whether there is any physiological difference between these two egg forms. It is possible that one exhibits a greater degree of dormancy than the other, as exhibited by other invertebrate groups. Such a situation is known to occur in some copepods, with the result that eggs that have been dormant for long periods (and screened from the selection pressures acting on the active population) have the effect of retarding the rate of evolution of the general population (Hairston and DeStasio, 1988).

An intermediate stage in the development of parthenogenesis from amphimictic populations (pseudogamy) has been observed in some nematodes (Nicholas, 1975). In these pseudogamous species, the spermatozoa trigger the cleavage divisions of the oocyte, but do not contribute any genetic material to the zygote. If the development of parthenogenesis in tardigrades has mirrored that in nematodes (Chapter 5), some tardigrades might be expected to exhibit pseudogamy. However, this intermediate mode of reproduction has not been recorded in tardigrades.

Structure

Although tardigrades have been observed by microscopists for a long time, new details of tardigrade anatomy are continually being published as new species are described and new techniques are employed. It can be seen, from the references cited in Chapter 4, that electron microscopy has added much to the body of knowledge generated by optical microscopic techniques. Other histochemical and microscopic techniques, such as confocal laser microscopy, will also help in the understanding of the structure and function of the internal structures in living tardigrades.

Ultrastructural details of structures, such as the sensillae and excretory organs, will help to explain the physiology and ecology of tardigarde species (and may also shed light on the phylogenetic relationships of the animals). Some anatomical observations made using optical apparatus, such as those of embryos by Marcus (1928; 1929*a*), have never been explored more fully using histochemical or electron microscopic techniques. Consequently, reviews such as this, must still rely on observations that are 60 years old. Comparative embryological studies may reveal a lot about the group, although the practical difficulties of such studies are formidable.

Many species are also poorly described and published material often lacks quantitative data. A review of the literature shows that the status of many species is frequently reappraised (e.g. Dastych, 1990; 1991; Villora-Moreno and

De Zio Grimaldi, 1993). Original descriptions (particularly of marine species) are often based on a small number of specimens so that the degree of intraspecific, ontogenic or sexual variation is unknown.

The British fauna

Recent observations of species new to the British Isles, such as *Ramazzottius varieornatus* and *Macrobiotus sandrae* from gutter sediment in Surrey (Bertolani and Kinchin, 1993), *Isohypsibius longiunguis* from a canal wall in Wiltshire (Marley et al., 1994) and *Echiniscus angolensis* from woodland moss in Slapton Ley Nature Reserve in South Devon (I.M. Kinchin, unpublished work), show that even limited studies from restricted areas may reveal new information. It is clear that further studies are likely to reward the observer with interesting results. It is also apparent that the published ranges of some species are not necessarily comprehensive: before the recent observation from southern England, *Echiniscus angolensis* had only previously been recorded from Africa and South America (Van Rompu et al., 1993; W. Maucci, personal communication; McInnes, 1994).

Many of the records of British tardigrades give very few details of the microhabitat sampled beyond statements like 'moss on a wall' or 'lichen on a tree trunk'. Details of the exposure and humidity type should be included, along with a grid reference so that later observers may sample the same microhabitat in order to record temporal changes. Notes on elevation and the pH of the underlying substratum (as given by Dastych, 1988) may also be of value. Understandably, few accounts describe the microhabitats from which no tardigrades are extracted and yet this information may provide clues to tardigrade habitat preferences. Some apparently ideal habitats contain no tardigrades (e.g. Kinchin, 1992*b*), but the reasons for this are unknown. Comments on the other meiofaunal groups present may also be helpful as they may be predatory upon tardigrades or provide prey items.

Given the availability of marine habitats around the British Isles, the British marine tardigrade fauna is represented by records of remarkably few species, reflecting the little attention that has been paid to this group. Morgan and O'Reilly (1988) list only 11 marine species from Scotland (*Halechiniscus perfectus, Echiniscoides sigismundi, Megastygarctides setoloso, Orzeliscus belopus* and seven species of *Batillipes*). This provides the observer with a great opportunity to make an original contribution to the body of knowledge on this group.

Acknowledgements

This book could not have been produced without the help of numerous colleagues. I would like to thank all those who have generously shared information with me and sent me numerous offprints, particularly: Clark Beasley (U.S.A.), Jan Bergström (Sweden), Roberto Bertolani (Italy), Vladimir Biserov (Russia), Hieronim Dastych (Germany), Ruth Dewel (U.S.A.), Diana Freckman (U.S.A.), Hartmut Greven (Germany), Susanna Grimaldi de Zio (Italy), Masamichi Ito (Japan), Reinhardt Kristensen (Denmark), Brian Mackness (Australia), Walter Maucci (Italy), Sandra McInnes (U.K.), Seung Yeo Moon (Korea), Clive Morgan (U.K.), Diane Nelson (U.S.A.), Giovanni Pilato and Maria Grazia Binda (Italy), David Pye (U.K.), Margherita Raineri (Italy), Lorena Rebecchi (Italy), Yves Séméria (France) and James Valentine (U.S.A.).

Many thanks to those who have given me permission to reproduce figures and photographs: Roberto Bertolani, Vladimir Biserov, Hieronim Dastych, R.P Esser, Hartmut Greven, Reinhardt Kristensen, Diane Nelson, Allen Lane (The Penguin Press), Gustav Fischer Verlag, Harvard University Press, Istituto Italiano di Idrobiologia, Paul Parey Verlag, Pergamon Press Ltd., Mucchi Editore, Royal Swedish Academy of Sciences, Scandinavian University Press, Zoological Society of Japan.

Particular thanks also to Roberto Bertolani, Hieronim Dastych, Diana Freckman, Anne Kinchin, Brian Mackness, Sandra McInnes and Terry O'Keeffe for their invaluable comments on the rough drafts of the chapters herein.

Finally, many thanks to Prof. Jim Green for his support over the years and to the biologists (and 'honorary' biologists) at The Royal Grammar School, Guildford for being themselves.

Glossary

Abiosis
see cryptobiosis

Acrocarpous
of mosses bearing the sporophyte at the apex of the main shoot, with further growth of the shoot by lateral innovations (cushion mosses)

Air plankton
microfauna transported passively on the wind in atmospheric moisture

Alae
cuticular extensions or 'wings' which greatly increase the surface area of some marine tardigrade species as an aid to dispersal. Can occur laterally or caudally. (Sing. ala)

Amphimixis
true sexual reproduction including the fusion of gametes from dioecious parents to produce a zygote (amphimictic)

Anabiosis
see cryptobiosis

Analogy
the possession of similar characteristics by different taxa, arising from developmental pathways unique to each group. The similarity is, therefore, a result of convergent evolution driven by similar environmental factors and not common genetic ancestry (analogous)

Anemochory
dispersal of organisms by wind

Anhydrobiont
an organism capable of using anhydrobiosis

Anhydrobiosis
a form of cryptobiosis initiated by desiccation (anhydrobiotic)

Animalcule
antiquated term used by pioneering microscopists of the 18th Century to describe motile micro-organisms. Tardigrades were sometimes refered to as *bear animalcules*

Anisotonic
of solutions with differing water potentials

Annelida
phylum of metamerically segmented roundworms

Anlage
the initial cells of an organ or special tissue in an embryo

Anoxybiosis
a form of cryptobiosis initiated by a reduction in oxygen tension

Apodeme
a skeletal process for the attachment of muscles

Apomixis
reproduction which occurs without fertilization and/or meiosis (apomictic)

Apomorphy
a character that is derived, and yet differs from the ancestral condition and may be of value in constructing a phylogeny (apomorphic)

Apophysis
a protruberance, contiguous with the underlying structure. (Plural apophyses)

Appressorium
a pad-like structure formed from a fungal hypha. This is pressed tightly onto the surface of the host. An infection peg grows from the appressorium through the cuticle of the host to initiate infection

Archenteron
cavity within an early embryo, communicating with the exterior via the blastopore. Ultimately becomes the gut cavity

Areolation
pattern of holes in the chorion of tardigrade egg shells between the ornamental projections

Arthropoda
a polyphyletic 'super-taxon' including three major extant evolutionary lines represented by the Crustacea, Chelicerata and Uniramia

Arthropodization
an evolutionary sequence of events stemming from the development of a stiffened exoskeleton which in turn leads to the development of other structures that are collectively recognized as characteristically arthropodan

Aschelminthes
a superphylum including the phyla Rotifera, Gastrotricha, Nematoda, Nematomorpha, Kinorhyncha, Loricifera, Priapulida and Chaetognatha

Automixis
the type of reproduction characterized by self-fertilizing hermaphrodites, male and female gametes derived from the same individual

Axenic
of cultures containing only one species (i.e. 'pure cultures'). Cultures of an organism with one other species (e.g. a food item) are monoxenic, those with two others are dixenic etc. Cultures of an undetermined mixture of organisms (e.g. a moss and its associated bryosystem) are described as xenic

Bauplan
from the German, meaning structural plan or design. Used in zoological texts to describe themes in animal architecture. (Plural *Baupläne*.) [See Brusca and Brusca (1990) for a discussion of the '*Bauplan* concept']

Benthos
fauna living on the floor of the sea (bottom-dwellers) (benthic)

Bilateria
a major zoological group that includes all metazoan phyla which exhibit bilateral body symmetry. Exceptions include the Porifera and Cnidaria which constitute the Radiata

Bi-sexual
of tardigrade populations consisting of both males and females

Blastopore
the opening formed by the invagination of cells in the early embryo, connecting its cavity (archenteron) to the outside. This pore is the site of the future mouth in the Protostomia and the anus in the Deuterostomia

Bound water
cellular water which is incorporated into hydrated molecules, as opposed to the 'free water' which provides an intra- and extracellular aqueous medium for metabolic activities

Bradytely
an unusually slow rate of morphological evolution which may result when a species avoids environmental selection pressures (by mechanisms such as cryptobiosis)

Browning reaction (Maillard reaction)
a sequence of chemical changes occurring without the involvement of enzymes during storage of products containing sugars and proteins, responsible for the surface colour change of bakery products and meat. The sequence begins with an aldol condensation reaction involving the carbonyl groups of the proteins and ends with the formation of furfural which produces the dark brown colouration which gives the reaction its name

Bryobiontic
of those animals which are usually only found in association with mosses, including all those which exhibit mimicry of mosses

Bryofauna
animal communities found living in a moss cushion (see also bryobiontic, bryophilous and bryoxenous)

Bryophagy
feeding upon bryophytes (bryophagous)

Bryophilous
of those animals which are often found in association with mosses, but are also common in other microhabitats

Bryophyte
plants belonging to the Division Bryophyta (mosses and liverworts)

Bryosystem
community of plants and animals living within the confines of a liverwort or moss cushion

Bryotelmata
water bodies held in the leaf axils and interstices of bryophytes. (Sing. bryotelm)

Bryoxenous
of animals which are only usually found in association with mosses during a specific phase of their life cycle (e.g. the larvae of some flies)

Calcicole
associated with an alkaline substratum

Calcifuge
associated with an acidic substratum

Carrying capacity
maximum number of organisms that can be supported in a given area

Caudal
of the tail region (e.g. the caudal limbs of tardigrades are the fourth pair of legs)

cf.
used to separate the generic and specific name in cases where the identity of the species is not certain, e.g. *Macrobiotus* cf. *hufelandi*, where a specimen belongs to the *hufelandi*-complex, but there is insufficient data (such as a lack of eggs) to confirm which species within the complex

Chelicerata
arthropod Phylum, including spiders (Aranae), mites (Acari) and other groups which feed using chelicerae. The body is divided into two parts: the prosoma (fused head and thorax) and the opisthosoma (abdomen)

Chorion
the superficial envelope or shell of invertebrate eggs (chorionic)

Cirrophore
cuticular structure supporting a cirrus

Clade
a group of species all descended from a common ancestor. The clade includes the ancestral species

Cloaca
terminal opening of the digestive tract into which the reproductive ducts also open. This arrangement is found in the Eutardigrada, whereas the Heterotardigrada feature separate anus and gonopore

Coelomate
animal that possess a coelom — an epithelium-lined space or body cavity between the body wall and the digestive tract in which other organs may be suspended (coelomic)

Convergence (convergent evolution)
evolutionary change which results in the modification of unrelated structures so that they become morphologically and functionally similar

Corticolous
(of lichens) growing on tree bark

Cosmopolitan
of euryplastic tardigrades which are common at all latitudes and altitudes. Those which are also common in urban/synanthropic habitats are termed holocosmopolitan; those which are not are termed hemicosmopolitan

Cretaceous
the most recent geological period of the Mesozoic Era, extending from 65–140 million years ago. The only known fossil tardigrades are of Cretaceous origin

Crossing over
a process whereby genes are exchanged between homologous chromosomes during meiosis

Crustacea
arthropod Phylum, including crabs and woodlice

Cryobiosis
a form of cryptobiosis initiated by a reduction in temperature

Cryptobiont
an organism capable of exhibiting cryptobiosis

Cryptobiosis
a reversible suspension of metabolism initiated by environmental extremes (cryptobiotic)

Cryptogam
collective term for non-flowering plants

Cyclomorphosis
a seasonal or cyclic change in morphology of an individual (cyclomorphic)

Deuterostomia
major zoological grouping of phyla which share a particular type of developmental pathway, includes the Echinodermata and Chordata. (See also Blastopore and Protostomia)

Diapause
a latent state triggered by hormonal changes within the organism, independent of short-term environmental changes. Common in some insects

Dimorphism
having two morphologically distinct forms. Sexual dimorphism — males and females which look different (dimorphic)

Dioecious
having the male and female gonads in separate organisms

Diploid
the number of chromosomes typical of the somatic cells of a species (twice the number found in the gametes. (See haploid)

Disseminules
structures used to promote the dispersal of a species, e.g. the seed of many flowering plants. The disseminules of tardigrades may be considered to be the tuns or eggs — particularly in terrestrial species employing anemochory, or the adults of marine species with alae to aid dispersal by ocean currents (hydrochory)

Diverticulum
a blind-ending tubular or sac-like outpushing from a cavity. (Plural, diverticula)

Ecdysis
moulting — periodic shedding of the cuticle

Echiniscid
member of the Family Echiniscidae (Heterotardigrada)

Echiniscoid
member of the Order Echiniscoidea (Heterotardigrada), including Families Oreelidae and Echiniscidae

Ectohydric
of mosses lacking internal water-conducting vessels. Water is conducted externally in channels on the surface of the plant. The plant stem acts as a wick, drawing water up from the substratum

Ediacara
a soft-bodied biotic assemblage characteristic of the Vendian Period. Also known as the Vendozoa. (See also Vendian)

Environment
the sum of the factors (both biotic and abiotic) acting upon an organism in its habitat

Epipelic
particularly of protoctists, especially diatoms, living on the surface of sand grains or other particles (epipelon)

Epiphyte
an organism living on the surface of plants (epiphytic)

Epizoite
an organism living on the surface of animals (epizoic)

Euryplastic
of species which can tolerate a wide variation in environmental conditions, often resulting in a wide geographical distribution. Can be used with a variety of suffixes to describe tolerance to variation in specific environmental factors: e.g. euryhydric (water); euryhaline (salinity); euryaerobic (oxygen). See also stenoplastic

Eutely
the phenomenon of cell constancy, when all members of a species exhibit a characteristic number of cells (as exhibited by some nematodes and certain tardigrade tissues). Growth is, therefore, achieved by cell expansion rather than mitosis

Exaptation
of any organ not evolved under natural selection for its current use. This may be because it performed a different function in ancestors or because it represented a non-functional part available for later co-option. This term is used in place of the older term 'pre-adaptation' which falsely implies fore-ordination

Excystation
emergence from an encysted condition

Exuvium
an empty, shed cuticle. (plural exuvia)

Ganglion
a solid mass of nervous tissue containing numerous cell bodies. (Plural ganglia)

Gibbosity
a term used by Italian tardigradologists which has been adopted some others to describe a hump in the cuticle of a eutardigrade

Gonochoristic
of dioecious animals possessing separate gonads used in amphimixis (gonochorism)

Gonopore
exit pore of the genital tract (separate from the anus in heterotardigrades), through which eggs are laid in females and sperm are passed from the male

Habitat
the place where an organism lives

Habitus
characteristic external appearance of the body

Haploid
the number of chromosomes typical of the gametes of a species

Hermaphrodite
an individual possessing both male and female reproductive systems

Hexapoda
uniramian sub-Phylum in which the adults typically possess six legs — insects. The adult body is typically divided into three: head (supporting the sensory structures such as antennae and compond eyes), thorax (supporting the legs and wings) and abdomen (containing the reproductive structures). The Hexapoda is the largest sub-group of the Arthropoda, consisting of perhaps one million species

Histolysis
the breakdown of tissues

Holoarctic
of the biogeographical region which includes the Nearctic region (i.e. North America) and the Palaearctic region (i.e. Europe, North Africa and North Asia)

Homology
having phenotypic characters that are of shared evolutionary derivation, i.e. a common genetic ancestry (homologous)

Hydrochory
dispersal of organisms by water (ocean currents, rivers or rain splashes which dislodge the animals from their substratum)

Hydrophobic
water-repellent, unwettable

Incertae sedis
Latin term meaning 'of uncertain position', used to describe taxa whose systematics are unclear. It is often considered preferable to describe a taxon as *incertae sedis* rather than allocate it to a dubious systematic position. (See also *Problematica*)

Instar
stage in the development of an arthropod, between two ecdyses (intermoult)

Interstices
spaces occurring between small structures, such as sand grains or moss stems. These spaces are frequently water-filled and often inhabited by small aquatic animals such as nematodes and tardigrades (interstitial)

Intrinsic rate of increase
the maximum number of offspring born per individual in a population

Isotonic
of solutions with identical water potentials

Iteroparity
reproducing repeatedly throughout adult life (iteroparous)

Karyological
concerning the chromosomes

Kinesis
a reaction to external stimulation in which the movements made are not related to the direction of the stimulus

K-selection
a life history strategy adapted to living at relatively constant population levels near the carrying capacity of a relatively stable habitat. Prone to density-dependent mortality

LD_{50}
test to determine the toxicity of a compound. LD_{50} is the dose that will kill 50% of the population. Also known as the median lethal dose (MLD)

Lobopod
poorly articulated limb typical of soft-bodied, arthropod-like animals (Pararthropoda) such as onychophorans and tardigrades. (The name lobopod is often used to describe the animals possessing these limbs, rather than just the limbs.) The taxon, Lobopodia, has been suggested by some authors to separate these animals from the eu-Arthropoda (insects, crustaceans etc.)

Lunule
a half-moon-shaped, sclerified structure sometimes occuring at the base of tardigrade claws. May or may not be dentate (toothed)

Macrobiotid
member of the Family Macrobiotidae (Eutardigrada)

Malpighian tubules (vasa Malpighi)
tubular excretory organs of insects, arachnids and myriapods. Terminology also used with less certainty for the excretory organs of eutardigrades

Meiofauna
animals small enough to pass through a sieve with a mesh size of 1 mm

Meiosis
process of cell division which results in two pairs of cells. Each cell has the haploid number of chromosomes. A process used in the production of gametes. (See mitosis)

Metamerism (metameric segmentation)
the repetition of a pattern of elements belongimg to each of the main organ systems of the body along the antero-posterior axis. Metamerism is most strongly developed in the annelids and arthropods. The repetition produces a series of segments or metameres

Metazoa
all multicellular animals

Micrometre (µm)
one-thousandth of a millimetre (10^{-6} m)

Mitosis
process of cell division which results in two identical cells each with the diploid number of chromosomes. A process used in growth. (See meiosis)

Monophyly
the derivation of a taxonomic group from a single ancestral lineage (monophyletic)

Morphometry (morphometric data)
quantitative morphological description, helpful in taxonomy, e.g. Pt ratios

Mucrone
small, cuticular tooth found in the buccal cavity of some tardigrade species

Multivoltine
of an organism that has several generations within a year. (See univoltine)

Myriapoda
uniramian sub-Phylum consisting of the Diplopoda (millipedes) and Chilopoda (centipedes)

Nanometre (nm)
one-millionth of a millimetre (10^{-9} m). 1 nm = 10 Ångstrom units

Neoteny
retention of juvenile characteristics in sexually mature individuals

Nematoda
an aschelminth Phylum of unsegmented roundworms (nematodes)

Onychophora
a taxon of uncertain affinities, sometimes described as 'walking worms'. The group is considered by some authors as a 'living fossil', representing the evolutionary link between annelids and arthropods. Some authors consider the Onychophora to be a Phylum, while others treat the group as a Class within the Arthropoda. Here the group is treated as a sub-Phylum of the Uniramia along with the Hexapoda, Myriapoda and (possibly) Tardigrada. (Onychophoran, *Peripatus*)

Oogenesis
process resulting in the formation of female gametes (ova) in the ovary

Organelle
a persistent structure within a cell. Usually membrane-bound and with specialized function, e.g. nucleus, mitochondrion

Osmobiosis
a form of cryptobiosis initiated by a decreased water potential

Parthenogen
an organism produced by parthenogenesis

Parthenogenesis
development of an individual from an egg that has not been fertilized by a male gamete

Phanerogam
collective name for flowering plants

Pharynx
the portion of the alimentary canal between the buccal cavity and intestine, characterized by a muscular bulb in tardigrades

Phoresy
a non-nutritional association between organisms involving transportation from one location to another by the host-carrier (phoretic)

Phylogeny
the history of the evolution of a species (as opposed to ontogeny — the history and development of an individual organism) (phylogenetic)

Phytotelmata
bodies of water held in the structure of flowering plants (e.g. leaf axils of teasel) and capable of supporting an aquatic microfaunal community. (Sing. phytotelm)

Placoid
cuticular thickening, occurring as a discrete rod, on the wall of the lumen in the pharynx of most tardigrades (usually differentiated into macro- and micro-placoids)

Plastron
a bubble or film of air trapped by projections, such as hairs or spines, held under water in communication with the gas-exchange surface. Functions as a store of air and a gill

Plesiomorphy
a primitive character which is similar to that exhibited by an ancestral species (plesiomorphic)

Pleurocarpous
mosses in which the female sex organs are borne on reduced lateral branches so that the gametophore bears several lateral sporophytes. (Feather mosses)

Poikilohydry
a physiological regime in which an organism's water content is largely controlled by environmental factors, such as humidity and insolation. Exhibited by mosses and tardigrades. Once dried, the organisms are usually resistant to further environmental deterioration. (See also anhydrobiosis.) (poikilohydric)

Polyphyly
the derivation of a taxonomic group from two or more ancestral lineages through convergent evolution (polyphyletic)

Polyploidy
having three or more times the haploid number of chromosomes. Polyploids are sterile when crossed with diploids and so reproduce by parthenogenesis or automixis

Problematica
a group of organisms of uncertain biological relationships. Term usually used with regard to fossil remains which often do not give enough information to permit the allocation of a definite systematic position (particularly in the Ediacara). Can also be used to describe living organisms such as tardigrades. (See also *incertae sedis*)

Proctodaeum
an invagination of ectoderm in an embryo, forming the anus or cloaca

Protandry
sequential development of gametes in hermaphrodites in which the male gametes develop before the female gametes. (See also protogyny)

Protoctista (Protista)
one of the five Kingdoms of organisms recognized by modern taxonomists. The group includes the eukaryotic micro-organisms and their immediate descendents, e.g. algae, slime moulds and protozoa

Protogyny
sequential development of gametes in hermaphrodites in which the female gametes develop before the male gametes. (See also protandry)

Protologue
the whole of the verbal and illustrative material associated with the description of a new taxon at its place of first valid publication

Protostomia
major zoological grouping of phyla which share a particular type of developmental pathway, includes the Arthropoda and Annelida. (See also Blastopore and Deuterostomia)

Psammolittoral
of the psammon of the littoral zone (sandy beach)

Psammon
fauna found living in sand (psammic)

Pseudogamy
reproduction in which the male gamete triggers the development of the female gamete, but does not contribute any genetic material to the zygote. An intermediate between parthenogenesis and amphimixis

Pt ratio
ratio of the size of various morphological characters with respect to the length of the buccal tube, e.g. Pt body length and Pt stylet insertion

Quiescence
a state of reduced metabolism, such as hibernation in mammals or encystment in tardigrades

r-selection
a life history strategy adapted to maximizing the rate of increase of a population in an unstable habitat. Prone to density-independent mortality. (See intrinsic rate of increase)

Saxicolous
(of lichens) growing on rock. (See also epilithic)

Scapus
collar at the base of a cirrus, separating the cirrophore from the flagellum

SEM
scanning electron microscope *or* scanning electron micrograph. SEMs show detail of surface structures at high magnifications coupled with a great depth of focus

Semelparity
reproducing only once in a life-time (semelparous)

Sensu lato (s.l.)
Latin term meaning, 'in the broad sense', e.g. *Ramazzottius oberhaeuseri* (*s.l.*) describing specimens that have been generally described as *R. oberhaeuseri*, but which may belong to other species now assigned to the *oberhaeuseri*-complex. At present, it may not be possible to re-examine all the specimens that were earlier referred to as *R. oberhaeuseri* to see if they are specimens of other species since described within the complex. Rather than ignore earlier observations, and to avoid perpetuating a possible misidentification, it is preferable to refer to the species *sensu lato*. (See cf.)

Sensu stricto (s.str.)
Latin term meaning, 'in the strict sense', e.g. *Macrobiotus hufelandi* (s.str.) describing the species recognized as *hufelandi* rather than a member of the *hufelandi*-complex. (See cf.)

Septulum
sclerified blades, alternating with and posterior to the microplacoids in the pharyngeal bulb of some *Diphascon* species. (Plural septula)

Simplex
tardigrade which has expelled its buccal apparatus to allow moulting

Species
the fundemental category of biological classification. Members of a species are physiologically able to interbreed to produce viable offspring. Species are genetically isolated from other species

Spermatogenesis
process resulting in the formation of male gametes — spermatozoa

Spur
a hook along the length of tardigrade claw, particularly in the echiniscids

Stenoplastic
of species with very precise and narrow environmental requirements which consequently restrict their distribution. Can be used with a variety of suffixes to describe intolerance to variation in specific environmental factors. (See euryplastic)

Sternite
a cuticular plate on the ventral surface of some heterotardigrades — typical of *Testechiniscus* spp. (Heterotardigrada)

Stomodaeum
an invagination of ectoderm in an embryo, forming the foregut and mouth

Stylet
sclerified buccal structure used in pairs by tardigrades to pierce the cells of their prey

Symbiosis
the living together of individuals of two different species. Often used instead of the more precise expression, mutualism, which implies an association in which both partners benefit (symbiotic, mutualistic). In its broadest sense, symbiosis includes other types of association, such as parasitism

Symphoriont
an organism involved in a phoretic relationship. (See phoresy)

Synanthropic
of habitats associated with human activity — walls, roofs etc.

Synapomorphy
the possession by two related lineages of the same phenotypic character derived from a different, but homologous, character in the ancestral lineage

Syngony
the production of male and female gametes in the same gonad (syngonic)

Systematics
study of the diversity of organisms and their evolutionary relationships — used in constructing a taxonomy

Taxis
a reaction to external stimulation where the movements made are related to the direction of the source of the stimulus. These may be positive (towards the source) or negative (away from the source)

Taxon
a single taxonomic unit. Arranged hierarchically, these include Kingdom, Phylum, Class, Order, Family, Genus, Species and various subdivisions (e.g. sub-Family, super-Family etc.) (Plural taxa)

Taxonomy
classifying organisms and dividing them into hierarchical groups or taxa

TEM
transmission electron microscope *or* transmission electron micrograph. TEMs show details of thin tissue sections with very high resolving power

Thelytoky
populations which are composed exclusively of females which reproduce by parthenogenesis (thelytokous)

Throttling
a constriction along the length of a pharyngeal placoid

Trehalose (α-D-glucopyranosyl-α-D-glucopyraniside)
a particularly stable non-reducing, disaccharide sugar ($C_{12}H_{22}O_{11}$) used as a membrane protectant during cryptobiosis in tardigrades. The sugar derives its name from a parasitic beetle, *Trehala manna*, the cocoons of which are rich in trehalose

Trilobita
an extinct arthropod Phylum (trilobites)

Tun
resistant stage adopted by cryptobiotic tardigrades. The volume of the body is reduced by 50% or more. This is achieved by an anterior–posterior contraction of the animal, multiple folding of the cuticle and retraction of the limbs. This is accompanied by an almost total loss of water by evaporation.

Ultrastructure
anatomical detail which can only be observed with the aid of an electron microscope, e.g. the structure of cell organelles (ultrastructural)

Uniramia
a monophyletic Phylum of the Arthropoda, including the sub-Phyla Hexapoda, Myriapoda, Onychophora and (possibly) the Tardigrada

Uni-sexual
of tardigrade populations consisting of one sex (females) only. Males are unknown for many species. (See thelytoky)

Univoltine
of an organism that has only one generation per year

Van der Land organ
structure of unknown function observed in the cirrophore of some arthrotardigrades

Vendian
the most recent period of the Precambrian Proterozoic Era dating from 550–650 million years ago. It seems probable that all the main evolutionary lines (leading to the extant phyla) had evolved by the end of the Vendian period. (See also Ediacara)

Vitellogenesis
process resulting in the development of the egg

Zoochory
dispersal of organisms by animals. This can be endo-zoochory (e.g. seeds that pass through a bird's gut) or ecto-zoochory (e.g. microfauna carried in debris on the feet or stuck to fur or feathers). Various prefixes can be used to indicate dispersal by particular animal groups, e.g. ornithochory, dispersal by birds

Zygote
diploid cell formed from the fusion of two haploid gametes. The zygote will develop into an embryo

References

Ashraf, M. (1971) Influence of environmental factors on Collembola. *Revue d'Ecologie et de Biologie du Sol,* **6**: 337–347

Baccetti, B. (1987) The evolution of the sperm cell in the phylum Tardigrada (Electron microscopy of Tardigrades. 5). In: Bertolani, R. (Ed.) *Biology of Tardigrades. Selected Symposia and Monographs* I. Collana U.Z.I., Modena, Mucchi Editore. pp. 87–91

Baccetti, B. and F. Rosati (1969) Electron microscopy on tardigrades I. Connective Tissue. *Journal of Submicroscopic Cytology,* **1**: 197–205

Baccetti, B. and F. Rosati (1971) Electron microscopy on tardigrades III. The Integument. *Journal of Ultrastructure Research,* **34**: 214–243

Baccetti, B., F. Rosati and G. Selmi (1971) Electron microscopy of tardigrades 4. The Spermatozoon. *Monitore Zoologico Italiano,* N.S., **5**: 231–240

Barrett, C.W. and R.G. Kimmel (1972) Effects of DDT on the density and diversity of tardigrades. *Proceedings of the Iowa Academy of Science,* **79**: 41–42

Barron, G.L. (1989) Host range studies for *Haptoglossa* and a new species, *Haptoglossa intermedia*. *Canadian Journal of Botany,* **67**: 1645–1648

Barron, G.L., C. Morikawa and M. Saikawa (1990) New Cephaliophora species capturing rotifers and tardigrades. *Canadian Journal of Botany,* **68**: 685–690

Baumann, H. (1961) Der Lebenslauf von *Hypsibius* (*H.*) *convergens* Urbanowicz (Tardigrada). *Zoologischer Anzeiger,* **165**: 123–128

Baumann, H. (1966) Lebenslauf und Lebensweise von *Hypsibius* (*H.*) *oberhaeuseri* Doyère (Tardigrada). *Veröeffentlichungen aus den Üebersee-Museum*, Bremen. Ser. A, **3**: 245–258

Beasley, C.W. (1988) Altitudinal distribution of Tardigrada of New Mexico with the description of a new species. *American Midland Naturalist,* **120**: 436–440

Bellido, A. and M. Bertrand (1981) *Echiniscoides travei* n. sp., un Tardigrade marin des iles Kerguelen (Heterotardigrada). *Bulletin museum national d'histoire naturelle*, Paris, Sect. A (Ser 4), **3**: 789–798.

Bennet-Clark, H.C. (1976) Mechanics of nematode feeding. In: Croll, N.A. (Ed.) *The Organization of Nematodes*. London, Academic Press. pp. 313–342

Bergström, J. (1986) Metazoan evolution — A new model. *Zoologica Scripta,* **15**: 189–200

Bergström, J. (1989) The origin of animal phyla and the new phylum Procoelomata. *Lethaia,* **22**: 259–269

Berlocher, S.H. (1982) Molecular systematics and taxonomic problems in the Tardigrada. In: Nelson, D.R. (Ed.) Proceedings of the third international symposium on the Tardigrada. East Tennessee State University Press, pp. 77–92

Bertolani, R. (1970a) Mitosi somatiche e costanza cellulare numerica nei Tardigradi. *Atti della Accademia Nazionale dei Lincei Rendiconti Classe di Scienze Fisiche Matematiche e Naturali,* **48**: 739–742

Bertolani, R. (1970b) Variabilité numerica cellulare in alcuni tessuti di Tardigradi. *Atti della Accademia Nazionale dei Lincei Rendiconti Classe di Scienze Fisiche Matematiche e Naturali,* **49**: 442–445

Bertolani, R. (1973a) Presenza di un biotipo partenogenetico e suo effetto sul rapporto-sessi in *Hypsibius oberhaeuseri* (Tardigrada). *Atti della Accademia Nazionale dei Lincei Rendiconti Classe di Scienze Fisiche Matematiche e Naturali,* Ser 8, **54**: 469–473

Bertolani, R. (1973b) Primo caso di una popolazione tetraploide nei Tardigradi. *Atti della Accademia Nazionale dei Lincei Rendiconti Classe di Scienze Fisiche Matematiche e Naturali,* Ser 8, **54**: 571–574

Bertolani, R. (1975) Citology and systematics in Tardigrada. In: Higgins, R.P. (Ed.) Proceedings of the first international symposium on tardigrades. *Memorie dell' Istituto Italiano di Idrobiologia,* **32** (Suppl.): 17–35

Bertolani, R. (1976) Osservazioni cariologiche su *Isohypsibius augusti* (Murray, 1907) e *I. megalonyx* Thulin, 1928 (Tardigrada) e ridescrizione delle due specie. *Bolletino di Zoologia*, **43**: 221–234

Bertolani, R. (1979a) Parthenogenesis and cytotaxonomy in Itaquasconinae (Tardigrada). *Zeszyty Naukowe Uniwersytetu Jagiellonskiego, Prace Zoologiczne*, **25**: 9–18

Bertolani, R. (1979b) Hermaphroditism in tardigrades. *International Journal of Invertebrate Reproduction*, **1**: 67–71

Bertolani, R. (1981a) The taxonomic position of some eutardigrades. *Bolletino di Zoologia*, **48**: 197–203

Bertolani, R. (1981b) A new genus and five new species of Italian fresh-water tardigrades. *Bolletino del Museo Civico di Storia Naturaledi*, Verona, **8**: 249–254

Bertolani, R. (1982a) Cytology and reproductive mechanisms in tardigrades. In: Nelson, D.R. (Ed.) Proceedings of the third international symposium on the Tardigrada. East Tennessee State University Press, pp. 93–114

Bertolani, R. (1982b) *Tardigradi (Tardigrada)*. Guide per il riconoscimento delle specie animale delle acque interne italiane. pp. 1–104

Bertolani, R. (1983a) Tardigradi muscicoli delle dune costiere italiane, con descrizione di una nuova specie. *Atti della società toscana di scienze naturali memorie*, Serie B, 90: 139–148

Bertolani, R. (1983b) Tardigrada. In: Adiyodi, K.G. and R.G. Adiyodi (Eds.) Reproductive Biology of Invertebrates. Volume 1: Oogenesis, Oviposition and Oosorption. Chichester, John Wiley & Sons Ltd. pp. 431–441

Bertolani, R. (1983c) Tardigrada. In: Adiyodi, K.G. and R.G. Adiyodi (Eds.) Reproductive Biology of Invertebrates. Volume 2: Spermatogenesis and sperm function. Chichester, John Wiley & Sons Ltd. pp. 387–396

Bertolani, R. (1987a) Sexuality, reproduction and propagation in tardigrades. In: Bertolani, R. (Ed.) *Biology of Tardigrades*. Selected Symposia and Monographs 1. Collana U.Z.I., Modena, Mucchi Editore. pp. 93–101

Bertolani, R. (1987b) *Biology of Tardigrades*. (Proceedings of the fourth international symposium on tardigrades) Selected Symposia and Monographs 1. Collana U.Z.I., Modena, Mucchi Editore, pp. 1–380

Bertolani, R. (1990) Tardigrada. In: Adiyodi, K.G. and R.G. Adiyodi (Eds.) Reproductive Biology of Invertebrates. Volume 4b: Fertilization, Development and Parental care. Chichester, John Wiley & Sons Ltd. pp.49–60

Bertolani, R. (1992) Tardigrada. In: Adiyodi, K.G. and R.G. Adiyodi (Eds.) Reproductive Biology of Invertebrates.Volume V: Sexual differentiation and behaviour. Chichester, John Wiley & Sons Ltd. pp.255–266

Bertolani, R., S. Garagna, G.C. Manicardi and C.A. Redi (1987) *Macrobiotus pseudohufelandi* Iharos as a model for cytotaxonomic study in populations of eutardigrades (Tardigrada). *Experientia*, (Basel) **43**: 210–213

Bertolani, R., S. Grimaldi De Zio, M. D'Addabbo Gallo and M.R. Morone De Lucia (1984) Postembryonic development in heterotardigrades. *Monitore Zoologico Italiano*, (N.S.) **18**: 307–320.

Bertolani, R. and I.M. Kinchin (1993) A new species of *Ramazzottius* (Tardigrada, Hypsibiidae) in a rain gutter sediment from England. *Zoological Journal of the Linnean Society*, London **109**: 327–333.

Bertolani, R. and R.M. Kristensen (1987) New records of *Eohypsibius nadjae* Kristensen, 1982, and revision of the taxonomic position of two genera of Eutardigrada (Tardigrada). In: Bertolani, R. (Ed.). *Biology of Tardigrades*. Selected Symposia and Monographs 1. Collana U.Z.I., Modena, Mucchi Editore, pp.359–372

Bertolani, R. and V. Mambrini (1977) Analisi cariologica e morfologica di alcune popolazioni di *Macrobiotus hufelandi* della Valsesia. *Atti della Accademia Nazionale dei Lincei Rendiconti Classe di Scienze Fisiche Matematiche e Naturali*, Ser 8, **62**: 239–245

Bertolani, R. and G.C. Manicardi (1986) New cases of hermaphroditism in tardigrades. *International Journal of Invertebrate Reproduction and Development*, **9**: 363–366

Bertolani, R. and G. Pilato (1988) Struttura delle unghie nei Macrobiotidae e descrizione di *Murrayon* n. gen. (Eutardigrada). *Animalia*, Catania, **15**: 17–24.

Bertolani, R., G. Pilato and M.A. Sabatini (1983) *Macrobiotus joannae*, primo Macrobiotidae ermafrodito (Eutardigrada). *Animalia*, Catania, **10**: 327–333

Bertolani, R. and L. Rebecchi (1993) A revision of the *Macrobiotus hufelandi* group (Tardigrada, Macrobiotidae), with some observations on the taxonomic characters of eutardigrades. *Zoologica Scripta*, **22**: 127–152

Bertolani, R.,L. Rebecchi and G. Beccaccioli (1990) Dispersal of *Ramazzottius* and other tardigrades in relation to type of reproduction. *Invertebrate Reproduction and Development*, **18**: 153–157

Binda, M.G. and G. Pilato (1986) *Ramazzottius*, nuovo genere di Eutardigrado (Hypsibiidae). *Animalia*, Catania, **13**: 159–166

Binda, M.G. and G. Pilato (1990) Tardigradi di Terra del Fuoco e Magallanes. 1. *Milnesium brachyungue*, nuova specie di tardigrado Milnesiidae. *Animalia*, Catania, **17**: 105–110

Bird, A.F. and M.S. Buttrose (1974) Ultrastructural changes in the nematode *Anguina tritici*, associated with anhydrobiosis. *Journal of Ultrastructural Research*, **4B**: 177–189

Biserov, V.I. (1990a) On the revision of the genus *Macrobiotus*. The subgenus *Macrobiotus* s.st.: a new systematic status of the group *hufelandi* (Tardigrada, Macrobiotidae). Communication 1. *Zoologicheskii Zhurnal*, **69**: 5–17 (in Russian with English summary)

Biserov, V.I. (1990b) On the revision of the genus *Macrobiotus*. The subgenus *Macrobiotus* s.st. is a new systematic status of the *hufelandi* group (Tardigrada, Macrobiotidae). Communication 2. *Zoologicheskii Zhurnal*, **69**: 38–50 (in Russian with English summary)

Biserov, V.I. (1992) A new genus and three new species of tardigrades (Tardigrada: Eutardigrada) from the USSR. *Bolletino di Zoologia*, **59**: 95–103

Blom, H-J. (1972) Haltung und Zucht von Bärtierchen. *Mikrokosmos*, **32**: S. 335–337

Boisseau, J-P. (1957) Technique pour l'étude quantitative de la faune interstitielle des sables. *Comptes Rendus du Congrés des Sociétés Savantes de Paris et des Départements*, **1957**: 117–119

Brook, A.J. (1993) Mainly for pleasure; or the amateur microscopist as a contributor to science. *The Quekett Journal of Microscopy*, **37**: 1–6

Brusca, R.C. and G.J. Brusca (1990) *Invertebrates*. Sunderland, Massachusetts: Sinauer Associates Inc.

Budd, G. (1993) A Cambrian gilled lobopod from Greenland. *Nature (London)*, **364**: 709–711

Bussau, C. (1992) New deep-sea Tardigrada (Arthrotardigrada, Halechiniscidae) from a manganese nodule area of the eastern South Pacific. *Zoologica Scripta*, **21**: 79–91

Bussers, J.C. and C. Jeuniaux (1973) Chitinous cuticle and systematic position of Tardigrada. *Biochemical Systematics*, **1**: 77–78

Cantacuzène, A. (1951) Tardigrade marin noveau, commensal de *Limnoria lignorum* (Rathke). *Comptes Rendus de l'Academie des Sciences. Paris*, **232**: 1699–1700

Carroll, J.J. and D.R. Viglierchio (1981) On the transport of nematodes by the wind. *Journal of Nematology*, **13**: 476–483

Cathey, D.D., B.C. Parker, G.M. Simmons, W.H. Yonge and M.R. Van Brunt (1981) The microfauna of algal mats and artificial substrates in Southern Victoria Land lakes of Antarctica. *Hydrobiologia*, **85**: 3–16

Christenberry, D. and W.H. Mason (1979) Redescription of *Echiniscus virginicus* Riggin (Tardigrada) with notes on life history, range and geographic variation. *Journal of the Alabama Academy of Science*, **50**: 47–61

Clarke, K.U. (1979) Visceral anatomy and arthropod phylogeny. In: Gupta, A.P. (Ed.) *Arthropod Phylogeny*. Van Nostrand Reinhold, pp. 467–547

Clegg, J.S. (1964) The control of emergence and metabolism by external osmotic pressure and the role of free glycerol in developing cysts of *Artemia salina*. *Journal of Experimental Biology*, **41**: 879–892

Clegg, J.S. (1965) The origin of trehalose and its significance during the formation of encysted dormant embryos of *Artemia salina*. *Comparative Biochemistry and Physiology*, **14**: 135–143

Clegg, J.S. (1973) Do dried cryptobiotes have a metabolism? In: Crowe, J.H. and J.S. Clegg (Eds.) *Anhydrobiosis*. Stroudsburg, Pennsylvania: Dowden, Hutchinson and Ross, pp. 141–146

Cloudsley-Thompson, J.L. (1988) *Evolution and adaptation of terrestrial arthropods*. Berlin, Springer Verlag.

Colaço, C., S. Sen, M. Thangavelu, S. Pinder and B. Roser (1992) Extraordinary stability of enzymes dried in trehalose: simple molecular biology. *Bio/Technology*, **10**: 1007–1011

Collin, J. and R.M. May (1950) Réactions adaptives de tardigrades à des variations de salinité. *Bulletin de la Société de Zoologie France*, **75**: 184–187

Cooper, K.W. (1964) The first fossil tardigrade: *Beorn leggi* Cooper from Cretaceous amber. *Psyche*, **71**: 41–48

Corbet, S.A. and O.B. Lan (1974) Moss on a roof, and what lives in it. *Journal of Biological Education*, **8**: 153–160

Couch, J.N. (1945) Observations on the genus *Catenaria*. *Mycologia*, **37**: 163–193.

Crisp, D.J. and J. Hobart (1954) LXVIII.— A note on the habitat of the marine tardigrade *Echiniscoides sigismundi* (Schultze). *Annals and Magazine of Natural History*, **7**: 554–560

Crisp M. and R.M. Kristensen (1983) A new marine interstitial eutardigrade from east Greenland, with comments on habitat and biology. *Videnskabelige Meddeleser fra Dansk Naturhistorisk Foreng*, **144**: 99–114

Croll, N.A. (1970) *The behaviour of nematodes their activity, senses and responses.* London, Edward Arnold

Crowe, J.H. (1972) Evaporative water loss by tardigrades under controlled relative humidities. *Biological Bulletin of the marine biological laboratory, Woods Hole*, **142**: 407–416

Crowe, J.H. (1975) The physiology of cryptobiosis in tardigrades. In: Higgins, R.P. (Ed.) Proceedings of the first international symposium on tardigrades. *Memorie dell'Istituto Italiano di Idrobiologia*, **32**(Suppl.): 37–59

Crowe, J.H., J.F. Carpenter, L.M. Crowe and T.J. Anchordoguy (1990) Are freezing and dehydration similar stress vectors? A comparison of modes of interaction of stabilizing solutes with biomolecules. *Cryobiology*, **27**: 219–231

Crowe, J.H., L.M. Crowe, J.F. Carpenter and C. Aurell Wistrom (1987) Stabilisation of dry phospholipid bilayers and proteins by sugars. *Biochemical Journal*, **242**: 1–10

Crowe, J.H. L.M. Crowe and D. Chapman (1984) Preservation of membranes in anhydrobiotic organisms: the role of trehalose. *Science*, **223**: 701–703

Crowe, J.H. and R.P. Higgins (1967) The revival of *Macrobiotus areolatus* Murray (Tardigrada) from the cryptobiotic state. *Transactions of the American Microscopical Society*, **86**: 286–294

Crowe, J.H. and K.A.C. Madin (1974) Anhydrobiosis in tardigrades and nematodes. *Transactions of the American Microscopical Society*, **93**: 513–524

Crowe, J.H. and K.A.C. Madin (1975) Anhydrobiosis in nematodes: evaporative water loss and survival. *Journal of Experimental Zoology*, **193**: 323–334

Crowe, J.H., I.M. Newell and W.W. Thomson (1970) *Echiniscus viridis* (Tardigrada): fine structure of the cuticle. *Transactions of the American Microscipocal Society*, **89**: 316–325

Crowe, J.H., I.M. Newell and W.W. Thomson (1971) Cuticle formation in the tardigrades, *Macrobiotus areolatus* Murray. *Journal de Microscopie (Paris)*, **11**: 121–132

Crowe, J.H., S.J. O'Dell and D.A. Armstrong (1979) Anhydrobiosis in nematodes: permeability during rehydration. *Journal of Experimental Zoology*, **207**: 431–438

Čuček, M. (1985) Tardigradi Cerkniskega Jazera in Okolice. *Bioloski Vestnik*, Lubljana, **33**: 1–10

Cuénot, L. (1932) Tardigrades. *Faune de France*, **24**: 1–96

D'Addabbo Gallo, M., R.M. Morone De Lucia and S. De Zio Grimaldi (1989) Two new species of the genus *Styraconyx* (Tardigrada: Heterotardigrada). *Cahiers de Biologie Marine*, **30**: 17–33

D'Addabbo Gallo, M., S. De Zio Grimaldi, R.M. Morone De Lucia and A. Troccoli (1992) Halechiniscidae and Echiniscoididae from the Western Mediteranean Sea. (Tardigrada: Heterotardigrada). *Cahiers de Biologie Marine*, **33**: 299–318

Dastych, H. (1974) North Korean Tardigrada. *Acta Zoologica Cracoviensia*, **19**: 125–145

Dastych, H. (1979) Tardigrada from Afghanistan with a description of *Pseudechiniscus schrammi* sp. nov. *Bulletin de la Société des Amis des Sciences et Lettres de Poznan, Série D, Sciences Biologiques*, **19**: 99–108

Dastych, H. (1983) *Apodibius confusus* gen. n. sp. n., a new water-bear from Poland (Tardigrada). *Bulletin of the Polish Academy of Sciences*, Biology, **31**: 41–46

Dastych, H. (1984) The Tardigrada from Antarctic with descriptions of new species. *Acta Zoologica Cracoviensia*, **27**: 377–436

Dastych, H. (1985) West Spitsbergen Tardigrada. *Acta Zoologica Cracoviensia*, **28**: 169–214

Dastych, H. (1987a) Altitudinal distribution of Tardigrada in Poland. In: Bertolani, R. (Ed.). *Biology of Tardigrades*. Selected Symposia and Monographs 1. Collana U.Z.I., Modena, Mucchi Editore, pp. 169–176.

Dastych, H. (1987b) Two new species of Tardigrada from the Canadian Subarctic with some notes on sexual dimorphism in the family Echiniscidae. *Entomologische Mitteilungen aus dem Zoologischen Museum Hamburg*, **8**: 319–334

Dastych, H. (1988) *The Tardigrada of Poland*. Monografie Fauny Polski, **16**: 1–255

Dastych, H. (1990) *Isohypsibius sattleri* (Richters, 1902), a valid species (Tardigrada). *Senckenbergiana biologica*, **71**: 181–189

Dastych, H. (1991) Redescription of *Hypsibius antarcticus* (Richters, 1904), with some notes on *Hypsibius arcticus* (Murray, 1907) (Tardigrada). *Mitteilungen aus dem Hamburgischen Zoologischen Museum und Institut*, **88**: 141-159

Dastych, H. (1992) *Paradiphascon manningi* gen. n. sp. n., a new water-bear from South Africa, with the erecting of a new subfamily Diphasconinae (Tardigrada). *Mitteilungen aus dem Hamburgischen Zoologischen Museum und Institut*, **89**: 125–139

Dastych, H. (1993) A new Genus and four new species of semiterrestrial Water-Bears from South Africa (Tardigrada). *Mitteilungen aus dem Hamburgischen Zoologischen Museum und Institut*, **90**: 175–186

Davis, R.C. (1981) Structure and function of two Antarctic terrestrial moss communities. *Ecological Monographs*, **5**: 125–143

De Barros, R. (1936) *Itaquascon umbellinae* gen. nov. spec. nov. (Tardigrada, Macrobiotidae). *Zoologischer Anzeiger*, **128**: 106–109

De Barros, R. (1942) Tardigrados do Estado de Sao Paulo, Brasil. II. Genero 'Macrobiotus'. *Revista Brasileira de Biologia*, **2**: 373–386

Delle Cave, L. and A.M. Simonetta (1975) Notes on the morphology and taxonomic position of *Aysheaia* (Onychophora?) and of *Skania* (undetermined Phylum). *Monitore Zoologico Italiano*, N.S.,**9**: 67–81

Demeure, Y., G. Reversat, S.D. Van Gundy and D.W. Freckman (1978) The relationship between nematode reserves and their survival to desiccation. *Nematropica*, **8**: 7–8

Demeure, Y., D.W. Freckman and S.D. Van Gundy (1979) Anhydrobiotic coiling of nematodes in soil. *Journal of Nematology*, **11**: 189–195

Dewel, R.A. and W.H. Clark (1973a) Studies on the tardigrades. II. Fine structure of the pharynx of *Milnesium tardigradum* Doyère. *Tissue and Cell*, **5**: 147–159

Dewel, R.A. and W.H. Clark (1973b) Studies on the tardigrades. III. Fine structure of the esophagus of *Milnesium tardigradum* Doyère. *Tissue and Cell*, **5**: 161–169

Dewel, R.A. and W.C. Dewel (1979) Studies on the tardigrades. IV. Fine structure of the hindgut of *Milnesium tardigradum* Doyère. *Journal of Morphology*, **161**: 79–110

Dewel, R.A. and W.C. Dewel (1990) The fine structure of the zoospore of *Sorochytrium milnesiophthora*. *Canadian Journal of Botany*, **68**: 1968–1977

Dewel, R.A., W.C. Dewel and B.G. Roush (1992) Unusual cuticle-associated organs in the heterotardigrade, *Echiniscus viridissimus*. *Journal of Morphology*, **212**: 123–140

Dewel, R.A., J.D. Joines and J.J. Bond (1985) A new chytridiomycete parasitizing the tardigrade *Milnesium tardigradum*. *Canadian Journal of Botany*, **63**: 1525–1534

Dewel, R.A., D.R. Nelson and W.C. Dewel (1993) Tardigrada. In: Harrison, F.W. and M.E. Rice (Eds.) Microscopic Anatomy of Invertebrates. Volume 12: Onychophora, Chilopoda and Lesser Protostomata. Chichester, Wiley–Liss Inc. pp. 143–183

Dewel, W.C. and R.A. Dewel (1987) Study of a moss community containing *Milnesium tardigradum* parasitized by a chytridiomyetous fungus. In: Bertolani, R. (Ed.). *Biology of Tardigrades*. Selected Symposia and Monographs 1. Collana U.Z.I., Modena, Mucchi Editore, pp. 45–56

De Zio, S. and P. Grimaldi (1966) Ecological aspects of Tardigrada distribution in S. Adriatic beaches.*Veröffentlichungen des Instituts für Meeresforschung in Bremerhaven*, **2**: 87–94

Doncaster, C.C. and D.J. Hooper (1961) Nematodes attacked by protozoa and tardigrades. *Nematologica*, **6**: 333–335

Donner, J. (1966) *Rotifers*. London, Frederick Warne & Co.

Dreschler, G.L. (1951) An entomophthoraceous tardigrade parasite producing small conidia on propulsive cells in spicate heads. *Bulletin of the Torrey Botanical Club*, **78**: 183–200

Durante, M.V. and W. Maucci (1972) Descrizione di *Hypsibius* (*Isohyps.*) *basalovoi* sp. nov. e altre notizie su Tardigradi del Veronese. *Memorie del Museo Civico di Storia Naturale di Verona*, **20**: 275–281

Dzik, J. and G. Krumbiegel (1989) The oldest 'onychophoran' *Xenusion*: a link connecting phyla? *Lethaia*, **22**: 169–181

Esser, R.P. (1990) Tardigrades attacking nematodes. Florida Department of Agriculture and Consumer Services, Division of Plant Industry. Nematology Circular Number 177

Esser, R.P. and N.E. El-Gholl (1992) *Harposporium*, a fungus that parasitizes and kills nematodes utilizing conidia swallowed by or sticking to its prey. Florida Department of Agriculture and Consumer Services, Division of Plant Industry. Nematology Circular Number 200

Evans, W.A. (1982) Abundances of micrometazoans in three sandy beaches in the island area of western Lake Erie. *Ohio Journal of Science*, **82**: 246–251

Everitt, D.A. (1981) An ecological study of an Antarctic freshwater pool with particular reference to Tardigrada and Rotifera. *Hydrobiologia*, **83**: 225–237

Fleeger, J.W. and W.D. Hummon (1975) Distribution and abundance of soil Tardigrada in cultivated and uncultivated plots of an old field pasture. In: Higgins, R.P. (Ed.) Proceedings of the first international symposium on tardigrades. *Memorie dell'Istituto Italiano di Idrobiologia*, **32** (Suppl.): 93–112

Fox, I. and I. Garcia-Moll (1962) *Echiniscus molluscorum*, new Tardigrade from the feces of the land snail, *Bulimulus exilis* (Gmelin) in Puerto Rico. *Journal of Parasitology*, **48**: 177–181

Franceschi Crippa, T. and A. Lattes (1967) Analisi della variazione della lunghezza degli esemplari di una popolazione di *Macrobiotus hufelandii* Schultze in rapporto con l'esistenza di mute. *Bolletino dei Musei e degli Istituto Biologici dell' Universita di Genova*, **35**: 45–54

Franceschi, T., M.L. Loi and R. Pierantoni (1962-63) Risultati di una prima indagine ecologica condotta su popolazioni di Tardigradi. *Bolletino dei Musei e degli Istituti Biologici dell' Universita di Genova*, **32**: 69–93

Frank, J.H. and L.P. Lounibos (1983) *Phytotelmata: Terrestrial plants as hosts for aquatic insect communities*. Medford, New Jersey: Plexus Publishing Inc.

Gerson, U. (1969) Moss–arthropod associations. *Bryologist*, **72**: 495–500

Gerson, U. (1982) Bryophytes and invertebrates. In: Smith, A.J.E. (Ed.) *Bryophyte Ecology*. London, Chapman and Hall, pp. 291–332

Giere, O. (1993) *Meiobenthology; the microscopic fauna in aquatic sediments*. Berlin, Springer-Verlag

Gilbert, J.J. (1974) Dormancy in rotifers. *Transactions of the American Microscopical Society*, **93**: 490–513

Gilbert, O.L. (1989) *The ecology of urban habitats*. London, Chapman and Hall

Gjelstrup, P. (1979) Epiphytic cryptostigmatid mites on some beech- and birch-trees in Denmark. *Pedobiologia*, **19**: 1–8

Gould, S.J. and E.S. Vrba (1982) Exaptation — a missing term in the science of form. *Paleobiology*, **8**: 4–15

Greaves, P.M. (1989) An introduction to the study of tardigrades. *Microscopy (London)*, **36**: 230–239

Greaves, P.M. (1991) Notes on the tardigrade fauna of Surrey. *Microscopy (London)*, **36**: 549–556

Green, J. (1950) Habits of the marine tardigrade *Echiniscoides sigismundi*. *Nature (London)*, **166**: 153–154

Gressitt, J.L. (1966) Epizoic symbiosis: The Papuan weevil genus *Gymnopholus* (Leptopiinae) symbiotic with cryptogamic plants, oribatid mites, rotifers and nematodes. *Pacific Insects*, **8**: 221–280

Gressitt, J.L. (1969) Epizoic symbiosis. *Entomological News*, **80**: 1–5

Greven, H. (1972) Vergleichende Untersuchungen am Integument von Hetero- und Eutardigraden. *Zeitschrift für Zellforschung und mikroskopische anatomie*, **135**: 517–538

Greven, H. (1975) New results and considerations regarding the fine structure of the cuticle in tardigrades. In: Higgins, R.P. (Ed.) Proceedings of the first international symposium on tardigrades. *Memorie dell'Istituto Italiano di Idrobiologia,* **32** (Suppl.): 113–131

Greven, H. (1976) Some ultrastructural observations on the midgut epithelium of *Isohypsibius augusti* (Murray, 1907) (Eutardigrada). *Cell and Tissue Research,* **166**: 339-351

Greven, H. (1979) Notes on the structure of vasa Malpighi in the eutardigrade *Isohypsibius augusti* (Murray 1907). *Zeszyty Naukowe Uniwersytetu Jagiellonskiego,* Prace Zoologiczne, **25**: 87–95

Greven, H. (1980) *Die Bärtierchen.* Die Neue Brehm-Bucherei, Vol. 537. Ziemsen Verlag

Greven, H. (1982) *Homologues or analogues? A survey of some structural patterns in the Tardigrada.* In: Nelson, D.R. (Ed.) Proceedings of the third international symposium on the Tardigrada. East Tennessee State University Press, pp. 55–76

Greven, H. (1984) Tardigrada. In: Bereiter-Hahn, J., A.G. Matoltsy and K.S. Richardson (Eds.). *Biology of the integument, Volume 1: Invertebrates.* Berlin, Springer-Verlag. pp. 714–727

Greven, H. and W. Greven (1987) Observations on the permeability of the tardigrade cuticle using lead as an ionic tracer. In: Bertolani, R. (Ed.) *Biology of Tardigrades.* Selected Symposia and Monographs 1. Collana U.Z.I. Modena, Mucchi Editore. pp. 35–43

Greven, H. and G. Grohé (1975) Die Feinstruktur des Integuments und der Muskelantzstellen von *Echiniscoides sigismundi* (Heterotardigrada). *Helgoländer wissenschaftliche Meeresuntersuchungen,* **27**: 450–460

Greven, H. and W. Peters (1986) Localization of chitin in the cuticle of Tardigrada using wheat germ agglutinin-gold conjugate as a specific electron-dense marker. *Tissue and Cell,* **18**: 297–304

Grigarick, A.A., R.O. Schuster and E.C. Toftner (1973) Descriptive morphology of eggs of some species in the *Macrobiotus hufelandi* group (Tardigrada: Macrobiotidae). *Pan-Pacific Entomologist,* **49**: 258–263

Grigarick, A.A., R.O. Schuster and E.C. Toftner (1975) Morphogenesis of two species of *Echiniscus.* In: Higgins, R.P. (Ed.) Proceedings of the first international symposium on tardigrades. *Memorie dell'Istituto Italiano di Idrobiologia,* **32** (Suppl.): 133–152

Grimaldi De Zio, S. (1986) Evoluzione dei Tardigradi Marini. *Nova Thalassia,* **8** (Suppl.3): 449–460

Grimaldi De Zio, S. and M. D'Addabbo Gallo (1975) Postembryonal development and moults in *Batillipes pennaki* Marcus (Heterotardigrada). *Rivista di Biologia,* **68**: 243–274

Grimaldi De Zio, S., M. D'Addabbo Gallo and M.R. Morone De Lucia (1980) Osservazioni sullo sviluppo post-embrionale di *Florarctus hulingsi* Renaud-Mornant (Heterotardigrada). *Memorie di Biologia Marina e di Oceanografia,* (N.S.) **10** (Suppl.): 407

Grimaldi De Zio, S., M. D'Addabbo Gallo and M.R. Morone De Lucia (1982a) *Neostygarctus acanthophorus* n. gen. n. sp., nuovo Tardigrado marino del Mediterraneo. *Cahiers de Biologie Marine,* **23**: 319–323

Grimaldi De Zio, S., M. D'Addabbo Gallo and M.R.Morone DeLucia (1982b) Note sull'ecologia dei tardigradi marini (Heterotardigrada). *Bollettino dei Musei e degli Istituto Biologici dell' Universita di Genova,* **50** (Suppl.): 223–227

Grimaldi De Zio, S., M. D'Addabbo Gallo, M.R. Morone De Lucia, R. Vaccarella and P. Grimaldi (1982c) Quattro nuove specie di Halechiniscidae in due grotte sotto-marine dell'Italia meridionale (Tardigrada, Heterotardigrada). *Cahiers de Biologie Marine,* **23**: 415–426

Grimaldi De Zio, S., M. D'Addabbo Gallo and M.R. Morone De Lucia (1983) Marine tardigrades ecology. *Oebalia,* **9**: 15–31

Grimaldi De Zio, S., M. D'Addabbo Gallo and M.R. Morone De Lucia (1984) Relazione tra morfologia ed ecologia nei tardigradi marini (Heterotardigrada-Arthrotardigrada). *Cahiers de Biologie Marine,* **25**: 67–73

Grimaldi De Zio, S., M. D'Addabbo Gallo and M. R. Morone De Lucia (1987) Adaptive radiation and phylogenesis in marine Tardigrada and the establishment of Neostygarctidae, a new family of Heterotardigrada. *Bolletino di Zoologia,* **54**: 27–33

Grimaldi De Zio, S., M. D'Addabbo Gallo and M.R. Morone De Lucia (1990a) Revision of the genus *Halechiniscus* (Halechiniscidae, Arthrotardigrada). *Cahiers de Biologie Marine,* **31**: 271–279

Grimaldi De Zio, S., M. D'Addabbo Gallo, M.R. Morone De Lucia and A. Troccoli (1990b) New description of *Neostygarctus acanthophorus* (Tardigrada, Arthrotardigrada). *Cahiers de Biologie Marine*, **31**: 409–416

Grimaldi De Zio, S., M. D'Addabbo Gallo and M. R. Morone De Lucia (1992) *Neoarctus primigenius* n. g., n. sp., a new Stygarctidae of the Tyrrhenian Sea (Tardigrada, Arthrotardigrada). *Bolletino di Zoologia*, **59**: 309–313

Grohé, G. (1976) Das marine Bärtierchen *Echiniscoides sigismundi. Mikrokosmos*, **5**: 129 -132

Hairston, N.G. and B.T. DeStasio (1988) Rate of evolution slowed by a dormant propagule pool. *Nature (London)*, **336**: 239–242

Hallas, T.E. (1971) Notes on the marine *Hypsibius-stenostomus* complex, with a description of a new species (Tardigrada, Macrobiotidae). *Steenstrupia*, **1**: 201–206

Hallas, T.E. (1972) Some consequences of varying egg-size in Eutardigrada. *Videnskabelige Meddeleser fra Dansk Naturhistorisk Foreng*, **135**: 21–31

Hallas, T.E. (1975a) Interstitial water and Tardigrada in a moss cushion. *Annales Zoologici Fennici*, **12**: 255–259

Hallas, T.E. (1975b) A mechanical method for the extraction of Tardigrada. In: Higgins, R.P. (Ed.) Proceedings of the first international symposium on tardigrades. *Memorie dell'Istituto Italiano di Idrobiologia*, **32** (Suppl.): 153–158

Hallas, T.E. (1978) Habitat preference in terrestrial tardigrades. *Annales Zoologici Fennici*, **15**: 66–68

Hallas, T.E. and R.M. Kristensen (1982) Two new species of the tidal genus *Echiniscoides* from Rhode Island, U.S.A. (Echiniscoididae, Heterotardigrada). In: Nelson, D.R. (Ed.) Proceedings of the third international symposium on the Tardigrada. East Tennessee State University Press, pp. 179–192

Hallas, T.E. and G.W. Yeates (1972) Tardigrada of the soil and litter of a Danish beech forest. *Pedobiologia*, **12**: 287–304

Hammer, M. (1966) A few oribatid mites from Ram, Jordan. *Zoologischer Anzeiger*, **177**: 272–276

Hanson, J. and J. Lowy (1960) Structure and function of the contractile apparatus in the muscle of invertebrate animals. In: Bourne, G.H. (Ed.) *The Structure and Function of Muscle Volume 1*, pp. 265–335, New York, Academic Press.

Harris, R.P. (1972) The distribution and ecology of the interstitial meiofauna of a sandy beach at Whitsand Bay, East Cornwall. *Journal of the Marine Biological Association of the UK*, **52**: 1–18

Higgins, R.P. (1959) Life history of *Macrobiotus islandicus* Richters with notes on other tardigrades from Colorado. *Transactions of the American Microscopical Society*, **78**: 137–157

Higgins, R.P. (1975) Proceedings of the first international symposium on tardigrades. *Memorie dell'Istituto Italiano di Idrobiologia*, **32** (Suppl.): 1–469

Higgins, R.P. (1977a) Meiofauna survives in dry beach sand. In: Gerlach, S.A. Means of meiofauna dispersal. *Mikrofauna Meeresboden*, **61**: 89–103

Higgins, R.P. (1977b) Sediment entangled in *Enteromorpha* algae fouling ships. In: Gerlach, S.A. Means of meiofauna dispersal. *Mikrofauna Meeresboden*, **61**: 89–103

Hinton, H.E. (1969) Respiratory systems of insect egg shells. *Annual Review of Entomology*, **14**: 343–368

Hofmann, I. (1987) Habitat preference of the most frequent moss-living Tardigrada in the area of Giessen (Hessen). In: Bertolani, R. (Ed.) *Biology of Tardigrades*. Selected Symposia and Monographs, 1. Collana U.Z.I., Modena, Mucchi Editore, pp. 211–216

Hou, X. and J. Chen (1989) Early Cambrian arthropod-annelid intermediate sea animal, *Luolishania longicruris* (gen. nov.) from Chengjiang, Yunnan. *Acta Palaeontologica Sinica*, **28**: 207–213 (in Chinese with English summary)

Hutchinson, M.T. and M.T. Streu (1960) Tardigrades attacking nematodes. *Nematologica*, **5**: 149

Iharos, G. (1968) The scientific results of the Soil Zoological Expeditions to South America. 6. Eine neue Tardigraden-Gattung von mariner Verwandschaft aus dem chilenischen Altiplano. *Opuscula Zoologia*, Budapest, **8** (2): 357–361

Iharos, G. (1973) Angaben zur geographischen Verbreitung der Tardigraden. *Opuscula Zoologia*, Budapest, **12**: 73–86

Iharos. G. (1975) Summary of the results of forty years of research on Tardigrada. In: Higgins, R.P. (Ed.) Proceedings of the first international symposium on tardigrades. *Memorie dell'Istituto Italiano di Idrobiologia*, **32** (Suppl.) 159–169

Inglis, W.G. (1985) Evolutionary waves: patterns in the origins of animal phyla. *Australian Journal of Zoology*, **33**: 153–178

Ito, M. (1991) Taxonomic Study on the Eutardigrada from the Northern Slope of Mt. Fuji, Central Japan. I. Families Calohypsibiidae and Eohypsibiidae. *Proceedings of the Japanese Society of Systematic Zoology*, **45**: 30–43

Ito, M. and K. Tagami (1993) A new species of the genus *Isohypsibius* (Eutardigrada: Hypsibiidae) collected from municipal water supply in Japan. *Japanese Journal of Limnology*, **54**: 81–83

Janetschek, H. (1967) Arthropod ecology of South Victoria Land. In: Gressitt, J.L. (Ed.) *Entomology of Antarctica*. Washington, American Geophysical Union, pp. 205–293

Jennings, P.G. (1975) The Signy Island terrestrial reference sites: V. Oxygen uptake of *Macrobiotus furciger* J. Murray (Tardigrada). *British Antarctic Survey Bulletin*, **41 & 42**: 161–168

Jennings, P.G. (1976) The Tardigrada of Signy Island, South Orkney Islands, with a note on the Rotifera. *British Antarctic Survey Bulletin*, **44**: 1–25

Jennings, P.G. (1979) The Signy Island terrestrial reference sites: X. Population dynamics of Tardigrada and Rotifera. *British Antarctic Survey Bulletin*, **47**: 89–105

Kathman, R.D. and S.F. Cross (1991) Ecological distribution of moss-dwelling tardigrades on Vancouver Island, British Columbia, Canada. *Canadian Journal of Zoology*, **69**: 122–129

Kathman, R.D. and D.R. Nelson (1987) Population trends in the aquatic tardigrade *Pseudobiotus augusti* (Murray). In: Bertolani, R. (Ed.) *Biology of Tardigrades*. Selected Symposia and Monographs, 1. Collana U.Z.I., Modena, Mucchi Editore, pp. 155–168

Keilin, D. (1959) The Leeuwenhoek lecture. The problem of anabiosis or latent life: history and current concept. *Proceedings of the Royal Society of London*, Series B, **150**: 149–191

Kinchin, I.M. (1985) Notes on the population dynamics of *Macrobiotus hufelandii* (Tardigrada). *Microscopy Bulletin*, **7**: 4–6.

Kinchin, I.M. (1987) The moss ecosystem. *School Science Review*, **68** (244): 499–503

Kinchin, I.M. (1989a) *Hypsibius anomalus* Ramazzotti (Tardigrada) from gutter sediment. *Microscopy (London)*, **36**: 240–244

Kinchin, I.M. (1989b) The moss fauna 2: Nematodes. *Journal of Biological Education*, **23**: 37–40

Kinchin, I.M. (1990a) The moss fauna 3: Arthropods. *Journal of Biological Education*, **24**: 93–99

Kinchin, I.M. (1990b) The cosmopolitan tardigrade *Milnesium tardigradum* Doyère: an observation from Northern Ireland. *Microscopy (London)*, **36**: 412–414

Kinchin, I.M. (1990c) Observations on the structure of *Ramazzottius* (with a checklist of British Eutardigrada). *Microscopy*, London, **36**: 475–482

Kinchin, I.M. (1992a) What is a tardigrade ? *Microscopy (London)*, **36**: 628–634

Kinchin, I.M. (1992b) An introduction to the invertebrate microfauna associated with mosses and lichens with observations from maritime lichens on the West coast of the British Isles. *Microscopy (London)*, **36**: 721–731

Kinchin, I.M. (1993) An observation on the body cavity cells of *Ramazzottius* (Hypsibiidae, Eutardigrada). *The Quekett Journal of Microscopy*, **37**: 52–55

Klekowski, R.Z. and K.W. Opalinski (1989) Oxygen consumption in Tardigrada from Spitsbergen. *Polar Biology*, **9**: 299–303.

Kristensen, R.M. (1976) On the fine structure of *Batillipes noerrevangi* Kristensen 1976. 1. Tegument and moulting cycle. *Zoologischer Anzeiger*, **197**: 129–150

Kristensen, R.M. (1977) On the marine genus *Styraconyx* (Tardigrada, Heterotardigrada, Halechiniscidae) with description of a new species from a warm spring on Disko Island, West Greenland. *Astarte*, **10**: 87–91

Kristensen, R.M. (1978) On the fine structure of *Batillipes noerrevangi* Kristensen 1978. 2. The muscle-attachments and the true cross-striated muscles. *Zoologischer Anzeiger*, **200**: 173–184

Kristensen, R.M. (1979) On the fine structure of *Batillipes noerrevangi* Kristensen 1978.

(Heterotardigrada). 3. Spermiogenesis. *Zeszyty Naukowe Uniwersytetu Jagiellonskiego*, Prace Zoologiczne, **25**: 97–105

Kristensen, R.M. (1980) Zur Biologie des marinen Heterotardigraden *Tetrakentron synaptae*. *Helgoländer Meeresuntersuchungen*, **34**: 165–177

Kristensen, R.M. (1981) Sense organs of two marine Arthrotardigrades (Heterotardigrada, Tardigrada). *Acta Zoologica*, **62** (1): 27–41

Kristensen, R.M. (1982a) The first record of cyclomorphosis in Tardigrada based on a new genus and species from Arctic meiobenthos. *Zeitschrift für Zoologische Systematik und Evolutionsforschung*, **20**: 249–270

Kristensen, R.M. (1982b) New aberrant eutardigrades from homothermic springs on Disko Island, West Greenland. In: Nelson, D.R. (Ed.) Proceedings of the third international symposium on the Tardigrada. East Tennessee State University Press, pp. 203–220

Kristensen, R.M. (1983) Loricifera, a new phylum with aschelminthes characters from the meiobenthos. *Zeitschrift für Zoologische Systematik und Evolutionsforschung*, **21**: 163–180

Kristensen, R.M. (1984) On the biology of *Wingstrandarctus corallinus* nov. gen. et spec., with notes on symbiotic bacteria in the subfamily Florarctinae (Arthrotardigrada). *Videnskabelige Meddeleser fra Dansk Naturhistorisk Foreng*, **145**: 201–218

Kristensen, R.M. (1987) Generic revision of the Echiniscidae (Heterotardigrada), with a discussion of the origins of the family. In: Bertolani, R. (Ed.) Biology of Tardigrades. Selected Symposium and Monographs, 1. Collana U.Z.I., Modena, Mucchi Editore, pp. 261–336

Kristensen, R.M. (1991) Loricifera–A general biological and phylogenetic overview. *Verhandlungen der Deutschen Zoologischen Gesellschaft*, **84**: 231–246

Kristensen, R.M. and T.E. Hallas (1980) The tidal genus *Echiniscoides* and its variability, with erection of Echiniscoididae fam.n. (Tardigrada). *Zoologica Scripta*, **9**: 113–127

Kristensen, R.M. and R.P. Higgins (1984a) A new Family of Arthrotardigrada (Tardigrada: Heterotardigrada) from the Atlantic Coast of Florida, U.S.A. *Transactions of the American Microscopical Society*, **103**: 295–311

Kristensen, R.M. and R.P. Higgins (1984b) Revision of *Styraconyx* (Tardigrada: Halechiniscidae), with descriptions of two new species from Disko Bay, West Greenland. *Smithsonian Contributions to Zoology*, **391**: 1–40

Kristensen, R.M. and R.P. Higgins (1989) Marine Tardigrada from the Southeastern United States Coastal Waters. I. *Paradoxipus orzeliscoides* n. gen., n. sp. (Arthrotardigrada: Halechiniscidae). *Transactions of the American Microscopical Society*, **108**: 262–282

Kristensen, R.M. and J. Renaud-Mornant (1983) Existence d'arthrotardigrades semi-benthiques de genres nouveaux de la sous-famille des Styraconyxinae subfam. nov. *Cahiers de Biologie Marine*, **24**: 337–353

Kronberg, I. (1983) *Ökologie der Schwarzen Zone im marinen Felslitoral: Monographie eines extremen Lebensraumes*. Thesis, Christian-Albrechts-Universität, Kiel. pp. 1–238

Lattes, A. (1975) Differences in the sculpture between adults and juveniles of *Echiniscus quadrispinosus*. A note on the importance of quantitative parameters in the systematics of the Echiniscidae. In: Higgins, R.P. (Ed.) Proceedings of the first international symposium on tardigrades. *Memorie dell'Istituto Italiano di Idrobiologia*, **32** (Suppl.): 171–176

Lawrence, P.N. and Z. Massoud (1973) Cuticle structures in the Collembola (Insecta). *Revue d'Ecologie et de Biologie du Sol*, **10**: 77–101

Lee, R.E. (1989) Insect cold-hardiness: to freeze or not to freeze. *Biological Science*, **39**: 308-313

Lindgren, E.W. (1971) Psammolittoral marine tardigrades from North Carolina and their conformity to worldwide zonation patterns. *Cahiers de Biologie Marine*, **12**: 481–496

Loomis, S.H., S.J. O'Dell and J.H. Crowe (1979) Anhydrobiosis in nematodes: inhibition of the browning reaction of reducing sugars with dry proteins. *Journal of Experimental Zoology*, **208**: 355–360

Loomis, S.H., K.A.C. Madin and J.H. Crowe (1980a) Anhydrobiosis in nematodes: biosynthesis of trehalose. *Journal of Experimental Zoology*, **211**: 311–320

Loomis, S.H., S.J. O'Dell and J.H. Crowe (1980b) Anhydrobiosis in nematodes: control and synthesis of trehalose during induction. *Journal of Experimental Zoology,* **211**: 321–330

Mackness, B.S. (1994) *A Bibliography of the Tardigrada.* Slow Walker Press, Melbourne: In Press

Madin, K.A.C. and J.H. Crowe (1975) Anhydrobiosis in nematodes: carbohydrate and lipid metabolism during dehydration. *Journal of Experimental Zoology,* **193**: 335–342

Maguire, B. Jr. (1959) Aquatic biotas of teasel waters. *Ecology,* **40**: 506

Maguire, B. Jr. (1963) The passive dispersal of small aquatic organisms and their colonization of isolated bodies of water. *Ecological Monographs,* **33**: 161–175

Manicardi, G.C. and R. Bertolani (1987) First contribution to the knowledge of alpine grassland tardigrades. In: Bertolani, R. (Ed.) *Biology of Tardigrades.* Selected Symposia and Monographs 1. Collana U.Z.I. Modena, Mucchi Editore, pp. 177–185

Manton, S.M. (1977) *The Arthropoda: habits, functional morphology and evolution.* Oxford, Clarendon Press

Marcus, E. (1928) Zur Embryologie der Tardigraden. *Verhandlungen der Deutschen Zoologischen Gessellschaft,* **32**: 134–146

Marcus, E. (1929a) Zur Embryologie der Tardigraden. *Zoologische Jahrbächer Abteilung für Anatomie und Ontogenie der Tiere,* **50**: 333–384

Marcus, E. (1929b) Tardigrada. In: Bronn, H.G. (Ed.). *Klassen und Ordnungen des Tierreichs.* Vol. 5, Section 4, Part 3: pp. 1–608

Marcus, E. (1936) Tardigrada. *Das Tierreich,* Berlin und Leipzig, **66**: 1–340

Marley, N.J., I.M. Kinchin and D.E. Wright (1994) A new addition to the Tardigrada fauna of Great Britain. II *Isohypsibius longiunguis* Pilato (Hypsibiinae, Hypsibiidae, Eutardigrada): In Press

Martinez, E.A. (1975) Marine meiofauna of a New York City beach, with particular reference to Tardigrada. *Estuarine and Coastal Marine Science,* **3**: 337–348

Massonneau, J. and R.M. May (1950) Le pigment des echinisciens. *Bulletin de la Societé de Zoologie (France),* **75**: 187–195

Masters, C.O. (1967) The biota of teasel waters as a science fair project. *Carolina Tips,* **30**: 1–2

Maucci, W. (1980) Analisi preliminare di alcuni dati statistici sulla ecologia dei Tardigradi muscicoli. *Bollettino del Museo Civico di Storia Naturale di Verona,* **7**: 1–47

Maucci, W. (1991) Tre nuove specie di Eutardigradi della Groenlandia meridionale. *Bollettino del Museo Civico di Storia Naturale di Verona,* **15**: 279–289

May, R.M. (1946) Cytologie des globules cavitaires actifs et dormants chez le tardigrade *Macrobiotus hufelandi* Schultze. *Archives d'Anatomie Microscopique et de Morphologie Experimentale,* **36**: 136–150

May, R.M. (1948) *La Vie des Tardigrades.* Paris, Gallimard

May, R.M. (1953) L'Evolution des tardigrades de la vie aquatic a la vie terrestre. *Bulletin Français de Pisciculture,* **25**: 93–100

McGinty, M.M. and R.P. Higgins (1968) Ontogenetic variation of taxonomic characters of two marine tardigrades with the description of *Batillipes bullacaudatus* n. sp. *Transactions of the American Microscopical Society,* **87**: 252–262

McInnes, S.J. (1991) Notes on tardigrades from the Pyrenees, including one new species. *Pedobiologia,* **35**: 11–26

McInnes, S.J. (1994) Zoogeographic distribution of terrestrial/freshwater tardigrades from current literature. *Journal of Natural History:* **28**: 257–352

McInnes, S.J. and J.C. Ellis-Evans (1987) Tardigrades from maritime Antarctic freshwater lakes. In: Bertolani, R. (Ed.) *Biology of Tardigrades.* Selected Symposia and Monographs, 1. Collana U.Z.I., Modena, Mucchi Editore, pp. 111–123

McInnes, S.J. and J.C. Ellis-Evans (1990) Micro-invertebrate community structure within a maritime Antarctic lake. *Proceedings of the National Institute of Polar Research Symposium on Polar Biology,* **3**: 179–189

McKenzie, K.G. (1983) On the origin of the Crustacea. *Memoirs of the Australian Museum,* **18**: 21–43.

McKirdy, D.J. (1975) *Batillipes* (Heterotardigrada): comparison of six species from Florida (USA) and a discussion of taxonomic characters within the genus. In: Higgins, R.P. (Ed.) Proceedings of the first international symposium on tardigrades. *Memorie dell'Istituto Italiano di Idrobiologia,* **32** (Suppl.): 177–223

Meininger, C.A. and P.D. Spatt (1988) Variations of tardigrade assemblages in dust-impacted Arctic mosses. *Arctic and Alpine Research,* **20**: 24–30

Meininger, C.A., G.W. Uetz and J.A. Snider (1985) Variation in epiphytic microcommunities (tardigrade–lichen–bryophyte assemblages) of the Cincinnati, Ohio area. *Urban Ecology,* **9**: 45–61

Mihelčič, F. (1950) Zur Physiologie und ökologie der Tardigraden. 1. Die Karotinoide ud ihre Bedeutung für Tardigraden. *Archivio Zoologico Italiano,* **35**: 349–360

Mihelčič, F. (1954/55) Zur ökologie der Tardigraden. *Zoologischer Anzeiger,* **153**: 250–257

Mihelčič, F. (1963) Können Tardigraden im Boden leben ? *Pedobiologia,* **2**: 96–101

Miller, J.D., P. Horne, H. Heatwole, W.R. Miller and L. Bridges (1988) A survey of the terrestrial Tardigrada of the Vestfold Hills, Antarctica. *Hydrobiologia,* **165**: 197–208

Mitchell, D. (1982) Some South-Eastern Tardigrades. *Transactions of the Kent Field Club* **9**: 9–12

Moon, S.Y., W. Kim and C.Y. Chang (1989) Freshwater tardigrades from Korea. *The Korean Journal of Systematic Zoology,* **5**: 159–171

Morgan, C.I. (1976) Studies on the British tardigrade fauna: Some zoogeographical and ecological notes. *Journal of Natural History,* **10**: 607–632

Morgan, C.I. (1977) Population dynamics of two species of Tardigrada, *Macrobiotus hufelandi* (Schultze) and *Echiniscus* (*Echiniscus*) *testudo* (Doyère), in roof moss from Swansea. *Journal of Animal Ecology,* **46**: 236–279

Morgan, C.I. and P.E. King (1976) *British Tardigrades. Tardigrada. Keys and notes for the identification of the species.* Synopses of the British Fauna (NS) No. 9. The Linnean Society/Academic Press

Morgan, C.I and D.J. Lampard (1986a) Supralittoral lichens as a habitat for tardigrades. *Glasgow Naturalist,* **21**: 127–138

Morgan, C.I and D.J. Lampard (1986b) The fauna of the Clyde sea area: Phylum Tardigrada. Occasional Publication No. 3. University Marine Biological Station, Millport, Isle of Cumbrae, pp. 1–46.

Morgan, C.I. and M. O'Reilly (1988) Additions to the Scottish tardigrade fauna, including a description of *Megastygarctides setoloso* new species, with a revised key for the identification of Scottish marine species. *Glasgow Naturalist,* **21**: 445–454

Müller, H.J. (1970) Formen der Dormanz bei Insekten. *Nova Acta Leopoldina,* **191**: 1–27

Murray, J. (1906) Scottish National Antarctic Expedition: Tardigrada of the South Orkneys. *Transactions of the Royal Society of Edinburgh,* **45**: 323–334

Murray, J. (1910) Tardigrada. *Report of the scientific investigations of the British Antarctic Expedition, 1907–1909 (E.H. Shackleton),* **1**: 81–185

Neel, J.K. (1948) A limnological investigation of the psammon in Douglas Lake, Michigan, with especial reference to shoal and shoreline dynamics. *Transactions of the American Microscopical Society,* **67**: 1–53

Nelson, D.R. (1975) Ecological distribution of tardigrades on Roan Mountain, Tennessee, North Carolina. In: Higgins, R.P. (Ed.) Proceedings of the first international symposium on tardigrades. *Memorie dell'Istituto Italiano di Idrobiologia,* **32** (Suppl.): 225–276

Nelson, D.R. (1982a) Developmental biology of the Tardigrada. In: Harrison, F.W. and R.R. Cowden (Eds.) Developmental Biology of Freshwater Invertebrates. New York, A.R. Liss. pp. 363–398

Nelson, D.R. (1982b) Proceedings of the third international symposium on tardigrades. August 3–6, 1980, Johnson City, Tennessee, U.S.A., East Tennessee State University Press, pp. 1–236

Nelson, D.R. (1991) Tardigrada. In: Thorp, J.H. and A.P. Covich (Eds.). Ecology and Classification of North American Freshwater Invertebrates. London, Academic Press. pp. 501 -521

Nelson, D.R., C.J. Kincer and T.C. Williams (1987) Effects of habitat disturbances on aquatic tardigrade populations. In: Bertolani, R. (Ed.) *Biology of Tardigrades.* Selected Symposia and Monographs, 1. Collana U.Z.I., Modena, Mucchi Editore, pp. 141–153

Nicholas, W. L. (1975) *The biology of free-living nematodes.* Oxford, Clarendon Press.

Noda, H. (1986) Seminal receptacle as a taxonomical character in the family Halechiniscidae (Arthrotardigrada). *Zoological Science,* **3**: 1111

Noda, H. (1987) A new species of marine Tardigrada of the genus *Florarctus* (Heterotardigrada, Halechiniscidae) from Japan. *Publication of the Seto Marine Biological Laboratory,* **32**: 323–328

Noda, H. (1993) Stygarctid tardigradev showing neoteny from Kuroshima island, Ryukyu Archipelago. *Zoological Science*, **10** (Suppl. 1): 174

Nörr, M. (1974) Hitzeresistenz bei Moosen. *Flora*, Jena, **163**: 388–397

Ottesen, P.S. and T. Meier (1990) Tardigrada from the Husvik area, South Georgia, sub-Antarctic. *Polar Research*, **8**: 291–294

Overgaard-Nielsen, C. (1948) Studies on the soil microfauna. II. The moss inhabiting nematodes and rotifers. *Naturvidenskabelige Skrifter Laerde Selsk Skrifter*, Aarhus, **1**: 1–98

Owen, R. (1855) Hunterian lecture on the comparative anatomy and physiology of invertebrate animals. Volume 1, Lecture No. 19 (second edition).

Parry, G.D. (1981) The meanings of r- and K-Selection. *Oecologia*, **48**: 260–264

Pennak, R.W. (1940) Ecology of the microscopic Metazoa inhabiting sandy beaches of some Wisconsin lakes. *Ecological Monographs*, **10**: 537–615

Pennak, R.W. (1951) Comparative ecology of the interstitial fauna of fresh-water and marine beaches. *Année Biologique*, **27**: 449–480

Pennak, R.W. (1989) Fresh-water Invertebrates of the United States (3rd Edition). New York, John Wiley and Sons Inc.

Pentecost, A. (1992) Travertine: life inside the rock. *Biologist*, **39**: 161–164

Perry, R.N. (1977) The effect of previous desiccation on the ability of the 4th stage larvae of *Ditylenchus dipsaci* to survive drying and control its rate of water loss. *Parasitology*, **75**: 215–231

Pianka, E.R. (1970) On r- and K-selection. *American Naturalist*, **104**: 592–597

Pigoń, A. and B. Węglarska (1953) The respiration of Tardigrada: A study of animal anabiosis. *Bulletin de l'Academie Polonaise des Sciences*, **1**: 69–72

Pigoń, A. and B. Węglarska (1955a) Anabiosis in Tardigrada. Metabolism and Humidity. *Bulletin de l'Academie Polonaise des Sciences*, **3**: 31–34

Pigoń, A. and B. Węglarska (1955b) Rate of metabolism in tardigrades during active life and anabiosis. *Nature* (London), **176**: 121–122

Pilato, G. (1969) Evoluzione e nuova sistemazione degli Eutardigrada. *Bolletino di Zoologia*, **36**: 327–345

Pilato, G. (1971) Tardigradi delle acque dolci siciliane. Nota prima. *Bollettino delle Sedute dell'Accademia Gioenia di Scienze Naturale*, Catania, **11**: 126–134

Pilato, G. (1972) Structure, intraspecific variability and systematic value of the buccal armature of eutardigrades. *Zeitschrift für Zoologische Systematik und Evolutionsforschung*, **10**: 65–78

Pilato, G. (1975) On the taxonomic criteria of the Eutardigrada. In: Higgins, R.P. (Ed.) Proceedings of the first international symposium on tardigrades. *Memorie dell'Istituto Italiano di Idrobiologia*, **32** (Suppl.): 277–304

Pilato, G. (1979) Correlations between cryptobiosis and other biological characteristics in some soil animals. *Bolletino di Zoologia*, **46**: 319–332

Pilato, G. (1981) Analisi di nuovi caratteri nello studio degli Eutardigrada, *Animalia*, Catania, **8**: 51–57

Pilato, G. (1982) The systematics of Eutardigrada: A comment. *Zeitschrift für Zoologische Systematik und Evolutionsforschung*, **20**: 271–284

Pilato, G. (1987) Revision of the genus *Diphascon* Plate, 1889, with remarks on the subfamily Itaquasconinae (Eutardigrada, Hypsibiidae). In: Bertolani, R. (Ed.) *Biology of Tardigrades*. Selected symposia and monographs 1. Collana U.Z.I. Modena, Mucchi Editore, pp. 337–357

Pilato, G. (1989) Phylogenesis and systematic arrangement of the family Calohypsibiidae Pilato, 1969 (Eutardigrada). *Zeitschrift für Zoologische Systematik und Evolutionsforschung*, **27**: 8–13

Pilato, G. (1992) *Mixibius*, nuovo genere di Hypsibiidae (Eutardigrada) *Animalia*, Catania **19**: 121–125

Pilato, G. and C.W. Beasley (1987) *Haplohexapodibius seductor* n. gen. n. sp. (Eutardigrada Calohypsibiidae) with remarks on the systematic position of the new genus. *Animalia*, Catania, **14**: 65–71

Pilato, G., R. Bertolani and M.G. Binda (1982) Studio degli *Isohypsibius* del gruppo *elegans* (Eutardigrada, Hypsibiidae) con descrizione di due nuove specie. *Animalia*, Catania, **9**: 185-198

Pilato, G. and M.G. Binda (1987a) *arascon schusteri* n. gen. n. sp. (Eutardigrada, Hypsibiidae, Itaquasconinae). *Animalia*, Catania, **14**: 91–97

Pilato, G. and M.G. Binda (1987b) *Richtersia*, nuovo genere di Macrobiotidae, e nuova definizione di *Adorybiotus* Maucci & Ramazzotti 1981 (Eutardigrada). *Animalia*, Catania, **14**: 147–152

Pilato, G. and M.G. Binda (1989) *Richtersius*, nuovo nome generico in sostituzione di *Richtersia* Pilato e Binda. 1987 (Eutardigrada). *Animalia*, Catania, **16**: 147–148

Pilato, G. and M.G. Binda (1990) Tardigradi Dell'Antartide. I. *Ramajendas*, nuovo genere di Eutardigrado. Nuova posizione sistematica di *Hypsibius renaudi* Ramazzotti, 1972 e descrizione di *Ramajendas frigidus* n. sp. *Animalia*, Catania, **17**: 61–71

Pilato, G. and M.G. Binda (1991) *Milnesium tetramellatum*, new species of Milnesiidae from Africa (Eutardigrada). *Tropical Zoology*, **4**: 103–106

Pilato, G, M.G. Binda and R. Catanzaro (1991) Remarks on some tardigrades of the African fauna with the description of three new species of *Macrobiotus* Schultze 1834. *Tropical Zoology*, **4**: 167–178

Pilato, G. and R. Catanzaro (1988) *Macroversum mirum* n. gen. n. sp. nuovo Eutardigrado (Macrobiotidae) dei Monti Nebrodi (Sicilia). *Animalia*, Catania, **15**: 175–180

Pohlad, B.R. and E.C. Bernard (1978) A new species of Entomophthorales parasitizing tardigrades. *Mycologia*, **70**: 130–139

Pollock, L.W. (1970a) Reproductive anatomy of some marine Heterotardigrada. *Transactions of the American Microscopical Society*, **89**: 308–316

Pollock, L.W. (1970b) Distribution and dynamics of interstitial Tardigrada at Woods Hole, Massachusetts, USA. *Ophelia*, **7**: 145–166

Pollock, L.W. (1971) On some British marine Tardigrada, including two new species of *Batillipes*. *Journal of the Marine Biological Association of the UK*, **51**: 93–103

Pollock, L. W. (1975a) Tardigrada. In: Giese, A.C. and J.S. Pearse (Eds.) Reproduction of marine invertebrates, Volume 2. Entoprocts and Lesser Coelomates. New York, Academic Press, pp. 43–54.

Pollock, L.W. (1975b) The role of three environmental factors in the distribution of the interstitial tardigrade *Batillipes mirus* Richters. In: Higgins, R.P. (Ed.) Proceedings of the first international symposium on tardigrades. *Memorie dell'Istituto Italiano di Idrobiologia*, **32** (Suppl.): 305–324

Pollock, L.W. (1975c) Observations on marine Heterotardigrada, including a new genus from the Western Atlantic Ocean. *Cahiers de Biologie Marine*, **16**: 121–132

Preston, C.M. and A.F. Bird (1987) Physiological and morphological changes associated with recovery from anabiosis in the dauer larva of the nematode *Anguina agrostis*. *Parasitology*, **95**: 125–133

Proctor, M.C.F. (1979) Structure and eco-physiological adaptations in bryophytes. In: Clarke, G.C.S. and J.G. Duckett (Eds.) *Bryophyte Systematics*. London, Academic Press, pp. 479–531

Proctor, M.C.F. (1982) Physiological ecology: Water relations, light and temperature responses, carbon balance. In: Smith, A.J.E. (Ed.) *Bryophyte Ecology*. London, Chapman and Hall, pp. 333–381

Proctor, M.C.F. (1984) Structure and ecological adaptation. In: Dyer, A.F. and J.G. Duckett (Eds.) *The Experimental Biology of Bryophytes*. London, Academic Press, pp. 9–37

Rahm, G. (1937) A new order of tardigrades from the hot springs of Japan. *Annotationes Zoologicae Japonenses*, **16**: 345–352

Raineri, M. (1982) Histochemical investigations of Tardigrada. I. Localization of cholinesterase activity. *Monitore Zoologico Italiano* (NS), **16**: 219–230

Raineri, M. (1985) Histochemical investigations of Tardigrada. 2. Alkaline phosphatase (ALP) and aminopeptidase (AMP) in the alimentary apparatus of Eutardigrada. *Monitore Zoologico Italiano* (NS), **19**: 47–67

Raineri, M. (1987) Histochemical investigations of Tardigrada. III. First evidence of neurosecretory cells in the cerebral ganglion of eutardigrades. In: Bertolani, R. (Ed.) *Biology of Tardigrades*. Selected symposia and monographs 1. Collana U.Z.I. Modena, Mucchi Editore, pp. 57–72

Ramazzotti, G. (1962) Il Phylum Tardigrada. *Memorie dell'Istituto Italiano di Idrobiologia*, **XVI**: 1–595

Ramazzotti, G. (1972) Il Phylum Tardigrada. (Seconda edizione aggiornata). *Memorie dell'Istituto Italiano di Idrobiologia*, **28**: 1–732

Ramazzotti, G. (1977) Note statistiche su una popolazione di *Macrobiotus areolatus* (Tardigrada).

Memorie dell'Istituto Italiano di Idrobiologia, **34**: 239–245

Ramazzotti, G. and W. Maucci (1982) A history of tardigrade taxonomy. In: Nelson, D.R. (Ed.) Proceedings of the third international symposium on the Tardigrada. August 3–6, 1980, Johnson City, Tennessee, U.S.A., East Tennessee State University Press, pp. 11–30

Ramazzotti, G. and W. Maucci (1983) Il Phylum Tardigrada (III edizione riveduta e aggiornata). *Memorie dell' Istituto Italiano di Idrobiologia*, **41**: 1–1012

Ramløv, H. and P. Westh (1992) Survival of the cryptobiotic eutardigrade *Adorybiotus coronifer* during cooling to -196° C: Effect of cooling rate, trehalose level, and short term acclimation. *Cryobiology*, **29**: 125–130

Ramsköld, L. and X. Hou (1991) New early Cambrian animal and onychophoran affinities of enigmatic metazoans. *Nature (London)*, **351**: 225–228

Rebecchi, L. (1991) Karyological analysis on *Macrobiotus pseudohufelandi* (Tardigrada, Macrobiotidae) and a new finding of a tetraploid population. *Caryologia*, **44**: 301–307

Rebecchi, L. and R. Bertolani (1988) New cases of parthenogenesis and polyploidy in the genus *Ramazzottius* (Tardigrada, Hypsibiidae) and a hypothesis concerning their origin. *Invertebrate Reproduction and Development*, **14**: 187–196

Rebecchi, L. and R. Bertolani (1992) Reproductive cycles in eutardigrades of freshwater and terrestrial habitats. Eighth International Meiofaunal Conference, University of Maryland, 9–14 August 1992 (Abstract)

Rebecchi, L. and A. Guidi (1991) First SEM studies on tardigrade spermatozoa. *Invertebrate Reproduction and Development*, **19**: 151–156

Redi, C.A. and S. Garagna (1987) Cytochemical evaluation of the nuclear DNA content as a tool for taxonomical studies in eutardigrades. In: Bertolani, R. (Ed.) *Biology of Tardigrades*. Selected symposia and monographs 1. Collana U.Z.I. Modena, Mucchi Editore, pp. 73–80

Renaud-Debyser, J. (1959) Sur quelques tardigrades du bassin d'Arcachon. *Vie et Milieu, Bulletin du Laboratoire Arago Université de Paris*, **10**: 135–146

Renaud-Debyser, J. (1964) Note sur la faune interstitielle du bassin d'Arcachon et description d'un gastrotriche nouveau. *Cahiers de Biologie Marine*, **5**: 111–123

Renaud-Mornant, J. (1970) *Parastygarctus sterreri* n. sp., Tardigrade marin nouveau de l'Adriatique. *Cahiers de Biologie Marine*, **11**: 355–360

Renaud-Mornant, J. (1974) Une nouvelle famille de tardigrades marins abyssaux: les Coronarctidae fam. nov. (Heterotardigrada). *Comptes Rendus de l'Academie des Sciences. Paris*, **278**: 3087-3090

Renaud-Mornant, J. (1979) Tardigrades marins de Madagascar. I. Halechiniscidae et Batillipedidae. *Bulletin museum national d'histoire naturelle*, Paris, Sect. A (Ser 4), **1**: 257-277

Renaud-Mornant, J. (1982a) Species diversity in marine Tardigrada. In: Nelson, D.R. (Ed.) Proceedings of the third international symposium on the Tardigrada. August 3–6, 1980, Johnson City, Tennessee, U.S.A., East Tennessee State University Press, pp. 149–178

Renaud-Mornant, J. (1982b) Sous-famille et genre nouveaux de Tardigrades marins (Arthrotardigrada). *Bulletin museum national d'histoire naturelle*, Paris, Sect. A (Ser 4), **4**: 89–94

Renaud-Mornant, J. (1983) Tardigrades abyssaux noveaux de la sous-famille des Euclavarctinae n. subfam. (Arthrotardigrada, Halechiniscidae). *Bulletin museum national d'histoire naturelle*, Paris, Sect. A (Ser 4), **5**: 201–219

Renaud-Mornant, J. (1984) Halechiniscidae (Heterotardigrada) de la campagne Benthedi, canal du Mozambique. *Bulletin museum national d'histoire naturelle*, Paris, Sect. A (Ser 4), **6**: 67-88

Renaud-Mornant, J. (1987) Halechiniscidae nouveaux de sables coralliens tropicaux (Tardigrada, Arthrotardigrada). *Bulletin museum national d'histoire naturelle*, Paris, Sect. A (Ser 4), **9**: 353–373

Renaud-Mornant, J. (1989) *Opydorscus*, un nouveau genre d'Orzeliscinae et sa signification phylogenique (Tardigrada, Arthrotardigrada). *Bulletin museum national d'histoire naturelle*, Paris, Sect. A (Ser 4), **11**: 763–771

Ricci, C.N. (1987) Ecology of bdelloids: how to be successful. *Hydrobiologia*, **147**: 117-127

Ricci, C.N. and G. Melone (1984) *Macrotrachela quadricornifera* (Rotifera, Bdelloidea); a SEM study on active and cryptobiotic animals. *Zoologica Scripta*, **13**: 195–200

Richardson, M. (1970) *Ballocephala verrucospora* sp. nov., parasitizing tardigrades. *Transactions of the British Mycological Society*, **55**: 307–309

Riggin, G.T. (1962) Tardigrada of Southwest Virginia: with the addition of a new marine species from Florida. *Virginia Agricultural Experiment Station Technical Bulletin*, **152**: 1–145

Robison, R.A. (1985) Affinities of *Aysheaia* (Onychophora) with description of a new Cambrian species. *Journal of Paleontology*, **59**: 226–235

Roff, D.A. (1992) *The Evolution of Life Histories: Theory and Analysis*. London, Chapman and Hall.

Rosati, F. (1968) Ricerche di microscopia elettronica sui Tardigradi, 2. I globuli cavitari. *Atti dell'Accademia dei Fisiocritici*, Sienna (Serie 13), **17**: 1439–1452

Roser, B. (1991a) Trehalose drying: a novel replacement for freeze-drying. *Biopharm*, **4**: 47–53

Roser, B. (1991b) Trehalose, a new approach to premium dried foods. *Trends in Food Science and Technology*, **7**: 166–169

Roser, B. and C. Colaço (1993) A sweeter way to fresher food. *New Scientist*, **1873**: 25–28

Rudolph, A.S. and J.H. Crowe (1985) Membrane stabilization during freezing: the role of two natural cryoprotectants, trehalose and proline. *Cryobiology*, **22**: 367–377

Ruppert, E.E. (1982) Comparative ultrastructure of the gastrotrich pharynx and the evolution of myoepithelial foreguts in Aschelminthes. *Zoomorphology*, **99**: 181–220

Saikawa, M. and M. Oyama (1992) Electron microscopy on infection of tardigrades by *Ballocephala sphaerospora* and *B. verrucospora* conidia. *Transactions of the Mycological Society of Japan*, **33**: 305–311

Saikawa, M., M. Oyama and K. Yamaguchi (1991) *Harposporium anguillulae* and *Haptoglossa intermedia* parasitizing tardigrades. *Transactions of the Mycological Society of Japan*, **32**: 501–508

Sayre, R.M. (1969) A method for culturing a predaceous tardigrade on the nematode, *Panagrellus redivivus*. *Transactions of the American Microscopical Society*, **88**: 266–274

Sayre, R.M. and S.W. Hwang (1975) Freezing and storing a tardigrade *Hypsibius myrops* in liquid nitrogen. *Cryobiology*, **12**: 568–569

Scheie, P.O. (1970) Environmental limits of cellular existence. *Journal of Theoretical Biology*, **28**: 315–325

Schuetz, G. (1987) A one-year study on the population dynamics of *Milnesium tardigradum* Doyère in the lichen *Xanthoria parietina* (L.) Th. Fr. In: Bertolani, R. (Ed.) *Biology of Tardigrades*. Selected Symposia and Monographs, 1. Collana U.Z.I., Modena, Mucchi Editore, pp. 217–228

Schulz, E. (1953) *Orzeliscus septentrionalis* n. sp. ein neuer mariner Tardigrad an der deutschen Nordseekuste. *Kieler Meeresforschung*, **9**: 288–292

Schulz, E. (1955) Studien an marinen Tardigraden. *Kieler Meeresforschung*, **11**: 74–79

Schuster, R.O., D.R. Nelson, A.A. Grigarick and D. Christenberry (1980). Systematic criteria of the Eutardigrada. *Transactions of the American Microscopical Society*, **99**: 284–303

Schuster, R.O., E.C. Toftner and A.A. Grigarick (1977) Tardigrada of Pope Beach, Lake Tahoe, California. *Wasmann Journal of Biology*, **35**: 115–136

Schwarz, A.M.J., J.D. Green, T.G.A. Green and R.D. Seppelt (1993) Invertebrates associated with moss communities at Canada Glacier, southern Victoria Land, Antarctica. *Polar Biology*, **13**: 157–162

Scott, G.A.M. (1982) Desert Bryophytes In: Smith, A.J.E. (Ed.) *Bryophyte Ecology*. London, Chapman and Hall, pp. 105–122

Séméria, Y. (1981) Recherches sur la faune urbaine et sub-urbaine des Tardigrades muscicoles et lichenicoles. 1. Nice-Ville. *Bulletin mensuel de la Société Linnéene de Lyon*, **50** (7): 231-237

Séméria, Y. (1982) Recherches sur la faune urbaine et sub-urbaine des Tardigrades muscicoles et lichenicoles. 2. L'espace sub-urbain: les hauter orientales de Nice-Ville. *Bulletin mensuel de la Société Linnéene de Lyon*, **51** (10): 315–328

Séméria, Y. (1993) Description d'une espace nouvelle de Tardigrade du Vénézuéla, *Ramazzottius edmondabouti* n. sp. (Eutardigrada Hypsibiidae). *Bulletin mensuel de la Société Linnéene de Lyon*, **62** (7): 215–216

Shaw, K. (1974) The fine structure of muscle cells and their attachments in the tardigrade *Macrobiotus hufelandi*. *Tissue and Cell*, **6**: 431–445

Simonetta, A.M. (1976) Remarks on the origins of the arthropods. *Memorie della Societá Toscana di Scienze naturale,* Ser B, **82**: 112–134

Simonetta, A.M. and S. Conway Morris (1991) *The early evolution of metazoa and the significance of problematic taxa.* Cambridge, Cambridge University Press.

Simonetta, A.M. and L. Delle Cave (1991) Early Palaeozoic arthropods and problems of arthropod phylogeny; with some notes on taxa of doubtful affinities. In: Simonetta, A.M. and S. Conway Morris (Eds.) *The early evolution of Metazoa and the significance of problematic taxa.* pp. 189–244. Cambridge, Cambridge University Press

Stearns, S.C. (1992) *The Evolution of Life Histories.* Oxford, Oxford University Press

Sterrer, W. (1973) Plate tectonics as a mechanism for dispersal and speciation in interstitial sand fauna. *Netherlands Journal of Sea Research,* **7**: 200–222

Steven, D.M. (1963) The dermal light sense. *Biological Reviews of the Cambridge Philosophical Society,* **38**: 204–240

Storch, V. (1993) *Pentastomida.* In: Harrison, F.W. and M.E. Rice (Eds.) Microscopic Anatomy of Invertebrates. Volume 12: Onychophora, Chilopoda and Lesser Protostomata. Chichester, Wiley-Liss Inc. pp. 115–142

Storey, K.B. and J.M. Storey (1990) Frozen and Alive. *Scientific American,* **263** (6): 62–67

Strickberger, M.W. (1990) *Evolution.* Boston, Jones and Bartlett

Sudzuki, M. (1972) An analysis of colonisation in freshwater microorganisms. II. Two simple experiments on dispersal by wind. *Japanese Journal of Ecology,* **22**: 222–225

Sugawara, H., K. Tanno, Y. Ohyama and H. Fukuda (1990) Freezing tolerance of *Macrobiotus harmsworthi* (Tardigrada) and *Plectus antarcticus* (Nematoda) in the Antarctic Region. *Antarctic Record (Nankyoku Shiryo),* **34**: 292–302 (in Japanese with English summary)

Thulin, G. (1928) Über die Phylogenie und das System der Tardigraden. *Hereditas,* **11**: 207–266.

Toftner, E.C., A.A. Grigarick and R.O. Schuster (1975) Analysis of scanning electron microscope images of Macrobiotus eggs. In: Higgins, R.P. (Ed.) Proceedings of the first international symposium on tardigrades. *Memorie dell'Istituto Italiano di Idrobiologia,* **32** (Suppl): 393–411

Triantaphyllou, A.C. and H. Hirschmann (1964) Reproduction in plant and soil nematodes. *Annual Review of Phytopathology,* **2**: 57–80

Tsurusaki, N. (1980) A new species of marine interstitial Tardigrada of the genus *Hypsibius* from Hokkaido, Northern Japan. *Annotationes Zoologicae Japonenses,* **53**: 280–284

Uhlig, G., H. Thiel and J.S. Gray (Eds.) (1973) The quantitative separation of meiofauna. A comparison of methods. *Helgoländer wissenschaftliche Meeresuntersuchungen,* **25**: 173–195

Usher, M.B. and H. Dastych (1987) Tardigrada from the maritime Antarctic. *British Antarctic Survey Bulletin,* **77**: 163–166

Utsugi, K. (1986) Urban tardigrades in Kyushu. *Zoological Science,* **3** (6): 1110.

Utsugi, K. and Y. Ohyama (1989) Antarctic Tardigrada. *Proceedings of the National Institute of Polar Research Symposium on Polar Biology,* **2**: 190–197.

Utsugi, K. and Y. Ohyama (1991) Antarctic Tardigrada II. Molodezhnaya and Mt. Riiser-Larsen areas. *Proceedings of the National Institute of Polar Research Symposium on Polar Biology,* **4**: 161–170.

Utsugi, K. and Y. Ohyama (1993) Antarctic Tardigrada III. Fildes Peninsula of King George Island. *Proceedings of the National Institute of Polar Research Symposium on Polar Biology,* **6**: 139–151

Valentine, J.W. (1989) Bilaterians of the Precambrian–Cambrian transition and the annelid–arthropod relationship. *Proceedings of the National Academy of Science, USA,* **86**: 2272–2275

Van der Land, J. (1964) A new peritrichous ciliate as a symphoriont on a tardigrade. *Zoologische Mededelingen,* Leiden, 39: 85–88

Van der Land, J. (1968) *Florarctus antillensis,* a new tardigrade from the coral sand of Curaçao. *Studies on the fauna of Curaçao and other Caribbean islands,* **25**: 140–146

Van der Land, J. (1975) The parasitic marine tardigrade *Tetrakentron synaptae.* In: Higgins, R.P. (Ed.) Proceedings of the first international symposium on tardigrades. *Memorie dell'Istituto Italiano di Idrobiologia,* **32** (Suppl.): 413–423

Van Rompu, E.A. and W.H. De Smet (1988) Some aquatic Tardigrada from Bjørnøya (Svalbard) (Norway). *Fauna Norvegica*, Series A, **9**: 31–36

Van Rompu, E.A. and W.H. De Smet (1991) Contribution to the fresh water Tardigrada from Barentsøya, Svalbard (78 degrees 30 minutes North). *Fauna Norvegica*, Series A, **12:** 29–39

Van Rompu, E.A., W.H. De Smet and J.M. Bafort (1992) Some freshwater tardigrades from the Kilimanjaro. *Natuurwetenschappelijk Tijdschrift*, **73**: 55–62

Vetter, J. (1990) The 'little bears' that evolutionary theory can't bear! *Creation Ex Nihilo*, **12**: 16–18

Villora-Moreno, S. and S. De Zio Grimaldi (1993) Redescription and ecology of *Batillipes phreaticus* Renaud-Debyser, 1959 (Arthrotardigrada, Batillipedidae) in the Gulf of Valencia (Western Mediterranean). *Cahiers de Biologie Marine*, **34**: 387–399

Volkmann, A. and H. Greven (1993) Ultrastructural localization of tyrosinase in the tardigrade cuticle. *Tissue and Cell*, **25**: 435–438

Von Brand, T. (1946) Anaerobiosis in invertebrates. *Biodynamica Monographs*, No. 4

Von Haffner, K. (1977) On the systematic position and the evolution of the Pentastomida as based on new comparative research. *Zoologischer Anzeiger*, **199**: 353–370

Wainberg, R.H. and W.D. Hummon (1981) Morphological variability of the tardigrade *Isohypsibius saltursus*. *Transactions of the American Microscopical Society*, **100**: 21–33

Wallwork, J.A. (1970) *Ecology of soil animals.* London, McGraw-Hill

Walz, B. (1973) Zur feinstruktur der muskelzellen des pharynx-bulbus von Tardigraden. *Zeitschrift für Zellforschung und mikroskopische anatomie*, **140**: 389–399

Walz, B. (1974) The fine structure of somatic muscles of Tardigrada. *Cell and Tissue Research*, **149**: 81–89

Walz, B. (1975a) Ultrastructure of muscle cells in *Macrobiotus hufelandi*. In: Higgins, R.P. (Ed.) Proceedings of the first international symposium on tardigrades. *Memorie dell'Istituto Italiano di Idrobiologia*, **32** (Suppl.): 425–444

Walz, B. (1975b) Modified ciliary structures in receptor cells of *Macrobiotus hufelandi* (Tardigrada). *Cytobiologie*, **11**: 181–185

Walz, B. (1978) Electron microscopic investigations of cephalic sense organs of the tardigrade *Macrobiotus hufelandi* C.A.S. Schultze. *Zoomorphologie*, **89**: 1–19

Walz, B. (1979a) The morphology of cells and cell organelles in the anhydrobiotic tardigrade *Macrobiotus hufelandi*. *Protoplasma*, **99**: 19–30

Walz, B. (1979b) Cephalic sense organs of *Tardigrada*. Current results and problems. *Zeszyty Naukowe Uniwersytetu Jagiellonskiego*, Prace Zoologiczne, **25**: 161–168

Walz, B. (1982) Molting in Tardigrada. A review including new results on cuticle formation in *Macrobiotus hufelandi*. In: Nelson, D.R. (Ed.) Proceedings of the third international symposium on the Tardigrada. August 3–6, 1980, Johnson City, Tennessee, U.S.A., East Tennessee State University Press, pp. 129–142

Węglarska, B. (1957) On the encystation in Tardigrada. *Zoologica Poloniae*, **8**: 315–325

Węglarska, B. (1975) Studies on the morphology of *Macrobiotus richtersi* Murray 1911. In: Higgins, R.P. (Ed.) Proceedings of the first international symposium on the tardigrades. *Memorie dell'Istituto Italiano di Idrobiologia*, **32** (Suppl.): 445–464

Węglarska, B. (1979a) Electron microscope study on previtellogenesis and vitellogenesis in *Macrobiotus richtersi* J. Murr. (Eutardigrada). *Zeszyty Naukowe Uniwersytetu Jagiellonskiego*, Prace Zoologiczne, **25**: 169–189

Węglarska, B. (1979b) Proceedings of the second international symposium on tardigrades. *Zeszyty Naukowe Uniwersytetu Jagiellonskiego*, Prace Zoologiczne, **25**: 1–197

Węglarska, B. (1980) Light and electron microscopic studies on the excretory system of *Macrobiotus richtersi*, Murray 1911 (Eutardigrada). *Cell and Tissue Research*, **207**: 171–182

Węglarska, B. (1987a) Studies on the excretory system of *Isohypsibius granulifer* Thulin (Eutardigrada). In: Bertolani, R. (Ed.) *Biology of Tardigrades.* Selected Symposia and Monographs, 1. Collana U.Z.I., Modena, Mucchi Editore, pp. 15–24

Węglarska, B. (1987b) Morphology and ultrastructure of excretory system in *Dactylobiotus dispar* (Murray)(Eutardigrada). In: Bertolani, R. (Ed.) *Biology of Tardigrades*. Selected Symposia and Monographs, 1. Collana U.Z.I., Modena, Mucchi Editore, pp.25–33

Węglarska, B. (1989a) Ultrastructure of tardigrade *Dactylobiotus dispar* (Murray 1907) cuticle. *Acta Biologica Cracoviensia*, **31**: 57–62

Węglarska, B. (1989b) Morphology of excretory organs in Eutardigrada. *Acta Biologica Cracoviensia*, **31**: 63–70

Westh, P. and R.M. Kristensen (1992) Ice formation in the freeze-tolerant eutardigrades *Adorybiotus coronifer* and *Amphibolus nebulosus* studied by differential scanning calorimetry. *Polar Biology*, **12**: 693–699

Westh, P., J. Kristiansen and A. Hvidt (1991) Ice-nucleating activity in the freeze-tolerant tardigrade *Adorybiotus coronifer*. *Comparative Biochemistry and Physiology [A]*, **73**: 621-626

Westh, P. and H. Ramløv (1991) Trehalose accumulation in the tardigrade *Adorybiotus coronifer* during anhydrobiosis. *Journal of Experimental Zoology*, **258**: 303–311

Wharton, D.A. (1980) Studies on the function of the oxyurid egg shell. *Parasitology*, **81**: 103–113

Wharton, D.A., C.M. Preston, J. Barrett and R.N. Perry (1988) Changes in cuticular permeability associated with recovery from anhydrobiosis in the plant parasitic nematode, *Ditylenchus dipsaci*. *Parasitology*, **97**: 317–330

Whitehead, A.G. and J.R. Hemming (1965) A comparison of some quantitative methods of extracting small vermiform nematodes from soil. *Annals of Applied Biology*, **55**: 25–38

Wieser, W. and J. Kanwisher (1959) Respiration and anaerobic survival in some seaweed-inhabiting invertebrates. *Biological Bulletin of the marine biological laboratory, Woods Hole*, **117**: 594–600

Willmer, P. (1990) *Invertebrate Relationships: Patterns in animal evolution*. Cambridge, Cambridge University Press

Wilson, E.O. (1992) *The Diversity of Life*. London, Allen Lane/The Penguin Press

Wolburg-Buchholz, K. and H. Greven, (1979) On the fine structure of the spermatozoon of Isohypsibius granulifer Thulin 1928 (Eutardigrada) with reference to its differentiation. *Zeszyty Naukowe Uniwersytetu Jagiellonskiego, Prace Zoologiczne*, **25**: 191–197

Womersley, C. (1981) Biochemical and physiological aspects of anhydrobiosis. *Comparative Biochemistry and Physiology*, **70B**: 669–678

Womersley, C. and L. Smith (1981) Anhydrobiosis in nematodes 1. The role of glycerol, myo-inisitol and trehalose during desiccation. *Comparative Biochemistry and Physiology*, **70B**: 579–586

Wright, J.C. (1987) *Anhydrobiosis in the Tardigrada*. DPhil Thesis, University of Oxford

Wright, J.C. (1988a) Structural correlates of permeability and tun formation in tardigrade cuticle: An image analysis study. *Journal of ultrastructure and molecular structure research*, **101**: 23–39

Wright, J.C. (1988b) The tardigrade cuticle. 1. Fine structure and distribution of lipids. *Tissue and Cell*, **20**: 745–758

Wright, J.C. (1989a) The tardigrade cuticle. 2. Evidence for a dehydration-dependent permeability barrier in the intracuticle. *Tissue and Cell*, **21**: 263–279

Wright, J.C. (1989b) Desiccation tolerance and water-retentive mechanisms in tardigrades. *Journal of Experimental Biology*, **142**: 267–292

Wright, J.C. (1991) The significance of four xeric parameters in the ecology of terrestrial Tardigrada. *Journal of Zoology (London)*, **224**: 59–77

Wright, J.C., P. Westh and H. Ramløv (1992) Cryptobiosis in Tardigrada. *Biological Reviews of the Cambridge Philosophical Society*, **67**: 1–29

Zullini, A. and E. Peretti (1986) Lead pollution and moss inhabiting nematodes of an industrial area. *Water, Air and Soil Pollution*, **27**: 403–410

Subject index

(Page numbers in boldface type indicate where figures are given.)

A

Acalciphil tardigrade species, 96
Actinarctus lyrophorus, **28**
Alae, 26, **28**
Altitudinal distribution of tardigrades, 99
Anhydrobiosis, 76
Annelid–uniramian evolutionary line, 12, **13**
Annex glands, 58
Anoxybiosis, 83
Apophyses, **42, 43**
Arthropodization, 10, 11
Arthropod phylogeny, 12
Arthropods, 113
Aysheaia spp. (LOBOPODIA), 7

B

Bacteria, 106, 107, **108**
Ballocephala sphaerospora (FUNGI), 106
Batillipes spp., **26, 70, 71, 92, 93, 136**
Batillipes digits, **70, 135**
Beorn leggi, 7
Body cavity cells, 54, **55**, 82, 85
British tardigrades, 145
Bryophagy in tardigrades, 109, 114
Buccal lamellae, **36**

C

Carnivory in tardigrades, 109
Calohypsibius ornatus, **35**
Cambrian fossils, **8**
Cephalic sensory appendages, 25, **26**
Chorionic plastron, 64

Cirri, 25, 51, 59
Cirrophore, 25, **26**
Clava, 25, 26, 51, 59
Claw gland, 67, **69**
Claws, 27, 33, **34, 35,** 124, **129**
Cloaca, 58
Coelomocytes, 54
Collection of specimens, 115
Collembolans, **111,** 113
Coronarctus tenellus, **94**
Cosmopolitan tardigrade species, 99
Cryobiosis, 82
Cryptobiosis, 75
Culturing tardigrades, 116
Cuticle, 45, **46, 47, 48**
Cuticle permeability, 78
Cuticle texture, **30**
Cuticular bars, **34,** 123
Cuticular plates, **29, 31**
Cyclomorphosis, 72, **73**
Cysts, 84, **85**

D

Dactylobiotus dispar, **134**
Dermal light sense, 140
Dimorphism, 58, 71
Diphascon scoticum, 95, 131, **133**
Dipsacus sylvestris (ANGIOSPERM), 89, **143**
Dispersal, 92, 142

E

Echiniscoides sigismundi, 75, 83, 84, 92, 135, **136, 137**
Echiniscus angolensis, 145
Echiniscus granulatus, **32,** 129

Echiniscus horningi, **107**
Echiniscus mauccii, **19**, **31**
Echiniscus merokensis, 30, **33**
Echiniscus molluscorum, 107
Echiniscus spiniger, **2**
Echiniscus tesselatus, **31**
Echiniscus testudo, 129, **130**
Echiniscus trisetosus, **31**
Echinursellus longiunguis, 18, **20**
Eggs, 62, **63**, 64, **66**, 72, **125**, **127**, 144
Embryo, **65**
Embryological development, 65
Embryonated egg, **66**
Eucalciphil tardigrade species, 96
Euryhydric tardigrade species, 96
Eutardigrada, 13
Eutardigrade systematics, 15, 18
Eutely, 67
Evolution of cryptobiosis, 75
Evolutionary relationships in heterotardigrades, **22**, **23**
Excretory organs, 52, **53**

F

Fertilization, 57
Freeze-tolerance, 82
Freshwater habitats, 89

G

Ganglia, 50
Gibbosities, 34
Gonads, **60**
Gonoporal dimorphism, 58
Gonopore, 58, **59**
Gymnophilus lichenifer (INSECTA), **142**

H

Halobiotus crispae, 53, 72, **73**
Harposporium sp. (FUNGI), 104, **105**
Hatching, 66
Hermaphrodite species, 57
Heterotardigrada, 13
Heterotardigrade systematics, 14, 18

Hoyer's medium, 117
Hydrophilous tardigrade species, 96
Hygrophilous tardigrade species, 78, 90, 96
Hypechiniscus gladiator, 30, **33**
Hypsibius antarcticus, 130, **131**
Hypsibius dujardini, 130, **131**

I

Ice-nucleating agents, 82
Instars, 70, **72**
International symposia, 4
Intestine, 39, **40**
Isohypsibius longiunguis, **85**, 145

K

Kerygmachela kierkegaardi (LOBOPODIA), 11
K-selection, 60

L

Leaf litter, 89
Lipids, 78, 82
Luolishania longicruris (LOBOPODIA), 7
Lunules, 34

M

Macrobiotus areolatus, 80, 126
Macrobiotus harmsworthi, 124, **126**
Macrobiotus hufelandi, **123**
Macrobiotus richtersi, 95, 126
Macrobiotus tonollii, **19**
Malpighian tubules, 52
Marine tardigrades, 90
Mating, 57
Mesocalciphil tardigrade species, 96
Mesocrista spitsbergense, 133
Mesotardigrada, 18
Metabolic relative temperature independence, 103
Midgut, 44
Milnesium tardigradum, **11**, **44**, 53, 59, 66, **112**, **127**

Subject Index

Minibiotus intermedius, **134**
Mites, **111**, 113
Moss, 87
Moulting, 67
Mucrones, **36**
Muscle attachment, **50**
Muscles, 48, **49**, **51**

N

Nematodes, 58, **105**, 110, **111**
Neoteny, 67
Nervous system, 50, **52**

O

Oesophagus, 44
Oligocalciphil tardigrade species, 96
Onychophora, 7, 12
Osmobiosis, 83
Ovotestis, 58
Oxygen consumption of tardigrades, 84

P

Parasitic relationships, 67, 90
Parastygarctus sp., 7, **26**, **28**
Parthenogenesis, 61
Pentastomida, 10, **11**
Peripatus sp. (ONYCHOPHORA), 49
Pharyngeal apparatus, **41**, 43, **45**, **68**, **132**
Pharyngeal bulb, **42**, **43**, **44**
Phototaxis, 140
Phytotelmata, 89, 143
Pigmentation, 29, 46, **122**
Placoids, 40, **43**
Poikilohydry, 88, 89
Polycalciphil tardigrade species, 96
Polyploidy, 61
Population densities, 102
Population studies on tardigrades, 100
Post-embryonic development, 66
Posteriodorsal apodeme, **41**, 44
Psammolittoral tardigrades, **93**

Psammon, 90
Pseudechiniscus novazeelandiae, **31**
Pseudobiotus megalonyx, 57, **65**
Pseudogamy, 144
Pt ratios, 120, **121**
Pylorus, **53**
Pyxidium tardigradum (PROTOZOA), 103, **104**

R

Ramazzottius anomalus, **122**
Ramazzottius cataphractus, **122**
Ramazzottius oberhaeuseri, 83, 128
Ramazzottius varieornatus, **42**, **69**, **78**, **121**, 145
Reproductive strategy, 98, 144
Richtersius coronifer, **81**, 83, 133, **134**
Rotifers, 109, **111**
r-selection, 60

S

Salivary glands, **40**, **44**, 67, **68**
Scapus, 25
Seminal receptacles, 26, **27**, 57, 58, 59
Sensillae, 51, **52**
Sensory fields, 52
Sex reversal, 59
Simplex, 67
Soil, 89
Sorochytrium milnesiophthora (FUNGI), 106
Specialised habitats, 142
Spermatozoa, 62
Stylets, 40, **43**

T

Tanarctus spp., **28**
Tardigrade behaviour, 139
Tardigrade classification, 14
Tardigrade phylogeny, 9
Testechiniscus spitsbergensis, **68**
Thermozodium esakii, 18, **20**
Tortula princeps (BRYOPHYTA), 88

Trehalose, 80, **81**, 82, 83, 141
Trophic relationships within a bryosystem, **110**
Tun, 77, **78**, **79**, **81**

U

Urban tardigrades, 143

V

Van der Land's organ, 26
Ventral organs, 53

X

X-bodies, 106
Xerophilous tardigrade species, 78, 89, 96

Z

Zonation of bryofauna, 93, **95**
Zonation of littoral tardigrades, **91**
Zonation of psammolittoral tardigrades, 91, **93**